Praise for

The Blood of Emmett Till

"Tim Tyson's genius as a historian, author, and social visionary informs his unique commitment to write truth to power authentically and fearlessly."

—Dr. Benjamin Chavis, former executive
director of the NAACP

"A critical book . . . [that] manages to turn the past into prophecy and demands that we do the one vital thing we aren't often enough asked to do with history: learn from it."

—Vann R. Newkirk II, *The Atlantic*

"What sets Tyson's book apart is the wide-angle lens he uses to examine the lynching, and the ugly parallels between past and present. . . . A terrific writer and storyteller, Tyson compels a closer look at a heinous crime and the consequential decisions, large and small, that made it a national issue."

—*Minneapolis Star Tribune*

"I couldn't stop reading Timothy Tyson's *The Blood of Emmett Till*. It is civil rights history that captivates the reader like a mystery novel."

—Patricia Bell-Scott, author of
The Firebrand and the First Lady

"*The Blood of Emmett Till* is a work critical not just to our understanding of something that happened in America in 1955 but of what happens in America here and now. It is a jolting and powerful book . . . swift-flying and meticulously researched."

—Leonard Pitts, *The Washington Post*

"An insightful, revealing and important new inquiry into the tragedy that mobilized and energized a generation of Americans to stand and fight against racial bigotry."

—Bryan Stevenson, *New York Times*
bestselling author of *Just Mercy*

"*The Blood of Emmett Till* unfolds like a movie, moving from scene to reconstructed scene, panning out to help the reader understand the racism and bigotry that crafted the citadel of white supremacy and focusing in on intimate exchanges imbued with meaning."
—Lawrence Jackson, *The Atlanta Journal-Constitution*

"Eloquent and outraged . . . A stunning success essential for our times."
—Nell Irvin Painter, author of *The History of White People*

"Tim Tyson has universalized the Emmett Till story to make it an American tragedy. His bracing, granular narrative provides fresh insight into the way race has informed and deformed our democratic institutions."
—Diane McWhorter, Pulitzer Prize–winning author of *Carry Me Home*

"No American historian working today captures the nuances of white supremacy and the ways in which it engulfs us all more convincingly than Tyson."
—Steve Nathans-Kelly, First of the Month

"An account of absorbing and sometimes horrific detail. Comprehensive in scope."
—*The New York Times*

"Tim Tyson's profound eloquence and groundbreaking evidence capture the cries of Emmett Till and the rise of a movement, and will call us to the cause of justice today."
—Rev. Dr. William J. Barber II, president of the North Carolina NAACP and author of *The Third Reconstruction: How a Moral Movement Is Overcoming the Politics of Division and Fear*

"From one of our finest civil rights historians comes this harrowing, brilliant, and crucial book. The full story of Emmett Till has never before been told. It will terrify you; it should. It will inspire you; it must."
—Jeff Sharlet, *New York Times* bestselling author of *The Family*

"Astonishingly relevant . . . at once thrilling and agonizing."
—*Jezebel*

"Tyson's powerful narrative sheds new light on the circumstances that led to the murder, makes the case that its influence stretches from the Montgomery bus boycott to the angry protests in Ferguson, Missouri—and argues that the country hasn't yet come to grips with the roots of any of the above."
—*Raleigh News & Observer*

"Tyson's profound conclusion moves the Emmett Till tragedy into the present time."

—*CounterPunch*

"Groundbreaking new evidence and Tyson's masterful prose make *The Blood of Emmett Till* a devastating indictment of America, both past and present."
—Danielle McGuire, author of *At the Dark End of the Street*

"Tyson gives us a history that challenges everything we thought we knew about Emmett Till."

—Crystal Feimster, author of *Southern Horrors*

"Tyson's meticulous and absorbing retelling of the events leading up to the horrific lynching in 1955 includes an admission from Till's accuser that some of her testimony was false."

—*The New York Times Book Review*

"Emotional and electric."

—*Toronto Star*

"When good and evil are evident, moral indignation comes easily, and readers might feel self-congratulatory, relieved that we are nothing like that anymore. We need historians like Timothy Tyson to break that spell for us."

—*Knoxville News Sentinel*

"*The Blood of Emmett Till* is less concerned with the historical cowardice of Bryant and the white men who effectively lynched Till, and much more invested in the bravery of Emmett Till's mother, Mamie, and the courage of the black activists who worked for voting rights and justice amidst the violent horror of life in Mississippi."

—*Yes! Weekly*

"Tyson's remarkable achievement is that each thread is explored in detail, backstories as well as main events, while he maintains a page-turning readability for what might seem a familiar tale. Cinematically engaging, harrowing, and poignant, Tyson's monumental work illuminates Emmett Till's murder and serves as a powerful reminder that certain stories in history merit frequent retelling."

—*Publishers Weekly* (starred review)

"This highly readable book is likely to remain the final account of the Till murder and trial and its impact in the United States and abroad."

—*Library Journal*

ALSO BY TIMOTHY B. TYSON

BLOOD DONE SIGN MY NAME: A TRUE STORY

RADIO FREE DIXIE: ROBERT F. WILLIAMS
AND THE ROOTS OF BLACK POWER

THE BLOOD OF
EMMETT TILL

TIMOTHY B. TYSON

SIMON & SCHUSTER PAPERBACKS

NEW YORK   LONDON   TORONTO   SYDNEY   NEW DELHI

Simon & Schuster Paperbacks
An Imprint of Simon & Schuster, Inc.
1230 Avenue of the Americas
New York, NY 10020

First Simon & Schuster trade paperback edition December 2017

For information about special discounts for bulk purchases,
please contact Simon & Schuster Special Sales at
1-866-506-1949 or business@simonandschuster.com.

The Simon & Schuster Speakers Bureau can bring authors to your live event.
For more information or to book an event contact the
Simon & Schuster Speakers Bureau at 1-866-248-3049 or
visit our website at www.simonspeakers.com.

Interior design by Ruth Lee-Mui

Manufactured in the United States of America

10   9

The Library of Congress has cataloged the hardcover edition as follows:
Names: Tyson, Timothy B., author.
Title: The blood of Emmett Till / Timothy B. Tyson.
Description: New York : Simon & Schuster, [2017] | Includes bibliographical
   references and index.
Identifiers: LCCN 2016021595 (print) | LCCN 2016023098 (ebook) | ISBN
   9781476714844 (hardcover) | ISBN 9781476714851 (pbk.) | ISBN 9781476714868
   (ebook) | ISBN 9781476714868 (E-Book)
Subjects: LCSH: Till, Emmett, 1941-1955. | Lynching—Mississippi—History—
   20th century. | African Americans—Crimes against—Mississippi. | Racism—
   Mississippi—History—20th century. | Trials (Murder)—Mississippi—
   Sumner. | Hate crimes—Mississippi. | United States—Race relations—
   History—20th century. | Mississippi—Race relations.
Classification: LCC HV6465.M7 T97 2017 (print) | LCC HV6465.M7
   (ebook) | DDC 364.1/34—dc23
LC record available at https://lccn.loc.gov/2016021595

ISBN 978-1-4767-1484-4
ISBN 978-1-4767-1485-1 (pbk)
ISBN 978-1-4767-1486-8 (ebook)

for my brother Vern

My name is being called on the road to freedom. I can hear the blood of Emmett Till as it calls from the ground. . . . When shall we go? Not tomorrow! Not at high noon! Now!

REVEREND SAMUEL WELLS, Albany, Georgia, 1962

CONTENTS

CONTENTS

1

NOTHING THAT
BOY DID

The older woman sipped her coffee. "I have thought and thought about everything about Emmett Till, the killing and the trial, telling who did what to who," she said.[1] Back when she was twenty-one and her name was Carolyn Bryant, the French newspaper *Aurore* dubbed the dark-haired young woman from the Mississippi Delta "a crossroads Marilyn Monroe."[2] News reporters from Detroit to Dakar never failed to sprinkle their stories about *l'affaire Till* with words like "comely" and "fetching" to describe her. William Bradford Huie, the Southern journalist and dealer in tales of the Till lynching, called her "one of the prettiest black-haired Irish women I ever saw in my life."[3] Almost eighty and still handsome, her hair now silver, the former Mrs. Roy Bryant served me a slice of pound cake, hesitated a little, and then murmured, seeming to speak to herself more than to me, "They're all dead now anyway." She placed her cup on the low glass table between us, and I waited.

For one epic moment half a century earlier, Carolyn Bryant's face had been familiar across the globe, forever attached to a crime of historic notoriety and symbolic power. The murder of Emmett Till was reported in one of the very first banner headlines of the civil rights era and launched the national coalition that fueled the modern civil rights movement. But she had never opened her door to a journalist or historian, let alone invited one for cake and coffee. Now she looked me in the eyes, trying hard to distinguish between fact and remembrance, and told me a story that I did not know.

The story I thought I knew began in 1955, fifty years earlier, when Carolyn Bryant was twenty-one and a fourteen-year-old black boy from Chicago walked into the Bryant's Grocery and Meat Market in a rural Mississippi Delta hamlet and offended her. Perhaps on a dare, the boy touched or even squeezed her hand when he exchanged money for candy, asked her for a date, and said goodbye when he left the store, tugged along by an older cousin. Few news writers who told the story of the black boy and the backwoods beauty failed to mention the "wolf whistle" that came next: when an angry Carolyn walked out to a car to retrieve the pistol under the seat, Till supposedly whistled at her.

The world knew this story only because of what happened a few days later: Carolyn's kinsmen, allegedly just her husband and brother-in-law, kidnapped and killed the boy and threw his body in the Tallahatchie River. That was supposed to be the end of it. Lesson taught. But a young fisherman found Till's corpse in the water, and a month later the world watched Roy Bryant and J. W. "Big" Milam stand trial for his murder.

I knew the painful territory well because when I was eleven years old in the small tobacco market town of Oxford, North Carolina, a friend's father and brothers beat and shot a young black man to death. His name was Henry Marrow, and the events leading up to his death had something in common with Till's. My father, a white Methodist minister, got mixed up in efforts to bring peace and justice to the community. We moved away that summer. But Oxford burned on in my memory, and I later went back and interviewed the man most responsible for Marrow's death. He told me, "That nigger committed suicide, coming in my store and wanting to four-letter-word my daughter-in-law." I also talked with many of those who had protested the murder by setting fire to the huge tobacco warehouses in downtown Oxford, as well as witnesses to the killing, townspeople, attorneys, and others. Seeking to understand what had happened in my own hometown made me a historian. I researched the case for years, on my way to a PhD in American history, and in 2004 published a book about Marrow's murder, what it meant for my hometown and my family, and how it revealed the workings of race in American history.[4] Carolyn Bryant Donham had read the book, which was why she decided to contact me and talk with me about the lynching of Emmett Till.

The killing of Henry Marrow occurred in 1970, fifteen years after the Till lynching, but unlike the Till case it never entered national or international awareness, even though many of the same themes were present. Like Till, Marrow had allegedly made a flirtatious remark to a young white woman at her family's small rural store. In Oxford, though, the town erupted into arson and violence, the fires visible for miles. An all-white jury, acting on what they doubtless perceived to be the values of the white community, acquitted both of the men charged in the case, even though the murder had occurred in public. What happened in Oxford in 1970 was a late-model lynching, in which white men killed a black man in the service of white supremacy. The all-white jury ratified the murder as a gesture of protest against public school integration, which had finally begun in Oxford, and underlying much of the white protest was fear and rage at the prospect of white and black children going to school together, which whites feared would lead to other forms of "race-mixing," even "miscegenation."

As in the Marrow case, many white people believed Till had violated this race-and-sex taboo and therefore had it coming. Many news reports asserted that Till had erred—in judgment, in behavior, in deed, and perhaps in thought. Without justifying the murder, a number of Southern newspapers argued that the boy was at least partially at fault. The most influential account of the lynching, Huie's 1956 presumptive tell-all, depicted a black boy who virtually committed suicide with his arrogant responses to his assailants. "Boastful, brash," Huie described Till. He "had a white girl's picture in his pocket and boasted of having screwed her," not just to friends, not just to Carolyn Bryant, but also to his killers: "That is why they took him out and killed him."[5] The story was told and retold in many ways, but a great many of them, from the virulently defensive accounts of Mississippi and its customs to the self-righteous screeds of Northern critics, noted that Till had been at the wrong place at the wrong time and made the wrong choices.

Until recently historians did not even have a transcript of the 1955 trial. It went missing soon after the trial ended, turning up briefly in the early 1960s but then destroyed in a basement flood. In September 2004 FBI agents located a faded "copy of a copy of a copy" in a private home in Biloxi, Mississippi. It took weeks for two clerks to transcribe the entire document, except for one missing page.[6] The transcript, finally released in 2007, allows us to compare the later recollections of witnesses and defendants with what they said fifty years earlier. It also reveals that Carolyn Bryant told an even harder-edged story in the courtroom, one that was difficult to square with the gentle woman sitting across from me at the coffee table.

Half a century earlier, above the witness stand in the Tallahatchie County Courthouse, two ceiling fans slowly churned the cigarette smoke. This was the stage on which the winner of beauty contests at two high schools starred as the fairest flower of Southern womanhood. She testified that Till had grabbed her hand forcefully across the candy counter, letting go only when she snatched it away. He asked her for a date, she said, chased her down the counter, blocked her path, and clutched her narrow waist tightly with both hands.

She told the court he said, "You needn't be afraid of me. [I've], well, ——with white women before." According to the transcript, the delicate young woman refused to utter the verb or even tell the court what letter of the alphabet it started with. She escaped Till's forceful grasp only with great difficulty, she said.[7] A month later one Mississippi newspaper insisted that the case should never have been called the "wolf whistle case." Instead, said the editors, it should have been called "an 'attempted rape' case."[8]

"Then this other nigger came in from the store and got him by the arm," Carolyn testified. "And he told him to come on and let's go. He had him by the arm and led him out." Then came an odd note in her tale, a note discordant with the claim of aborted assault: Till stopped in the doorway, "turned around and said, 'Goodbye.'"[9]

The defendants sat on the court's cane-bottom chairs in a room packed with more than two hundred white men and fifty or sixty African Americans who had been crowded into the last two rows and the small, segregated black press table. In his closing statement, John W. Whitten, counsel for the defendants, told the all-white, all-male jury, "I'm sure that every last Anglo-Saxon one of you has the courage to free these men, despite this [outside] pressure."[10]

Mamie Bradley,* Till's mother, was responsible for a good deal of that outside pressure on Mississippi's court system. Her brave decision to hold an open-casket funeral for her battered son touched off news stories across the globe. The resultant international outrage compelled the U.S. State Department to lament "the real and continuing damage to American foreign policy from such tragedies as the Emmett Till case."[11] Her willingness to travel anywhere to speak about the tragedy helped to fuel a huge protest movement that pulled together the elements of a national civil rights movement, beginning with the political and cultural power of black Chicago. The movement became the most important legacy of the story.[12] Her memoir of the case, *Death of Innocence*, published almost fifty years

*Mamie Carthan became Mamie Till after her marriage to Louis Till in 1940, which ended with his death in 1945. Mamie Till became Mamie Mallory after a brief remarriage in 1946. Her name changed to Bradley after another marriage in 1951. She was Mamie Bradley during most of the years covered by this book. She married one last time in 1957, becoming Mamie Till-Mobley, under which name she published her 2004 memoir. To avoid confusion, and also to depict her as a human being rather than an icon, I generally refer to her by her first name. No disrespect is intended. The same is true of Emmett Till and Carolyn Bryant.

after her son's murder, lets us see him as a human being, not merely the victim of one of the most notorious hate crimes in history.[13]

As I sat drinking her coffee and eating her pound cake, Carolyn Bryant Donham handed me a copy of the trial transcript and the manuscript of her unpublished memoir, "More than a Wolf Whistle: The Story of Carolyn Bryant Donham." I promised to deliver our interview and these documents to the appropriate archive, where future scholars would be able to use them. In her memoir she recounts the story she told at the trial using imagery from the classic Southern racist horror movie of the "Black Beast" rapist.[14] But about her testimony that Till had grabbed her around the waist and uttered obscenities, she now told me, "That part's not true."

A son of the South and the son of a minister, I have sat in countless such living rooms that had been cleaned for guests, Sunday clothes on, an unspoken deference running young to old, men to women, and, very often, dark skin to light. As a historian I have collected a lot of oral histories in the South and across all manner of social lines. Manners matter a great deal, and the personal questions that oral history requires are sometimes delicate. I was comfortable with the setting but rattled by her revelation, and I struggled to phrase my next question. If that part was not true, I asked, what did happen that evening decades earlier?

"I want to tell you," she said. "Honestly, I just don't remember. It was fifty years ago. You tell these stories for so long that they seem true, but that part is not true." Historians have long known about the complex reliability of oral history—of virtually all historical sources, for that matter—and the malleability of human memory, and her confession was in part a reflection of that. What does it mean when you remember something that you know never happened? She had pondered that question for many years, but never aloud in public or in an interview. When she finally told me the story of her life and starkly different and much larger tales of Emmett Till's death, it was the first time in half a century that she had uttered his name outside her family.

Not long afterward I had lunch in Jackson, Mississippi, with Jerry Mitchell, the brilliant journalist at the *Clarion-Ledger* whose sleuthing has solved several cold case civil rights–era murders. I talked with him about my efforts to write about the Till case, and he shared some thoughts of his own. A few days after our lunch a manila envelope with a Mississippi return address brought hard proof that "that part," as Carolyn had called the alleged assault, had never been true.

Mitchell had sent me copies of the handwritten notes of what Carolyn Bryant told her attorney on the day after Roy and J.W. were arrested in 1955. In this earliest recorded version of events, she charged only that Till had "insulted" her, not grabbed her, and certainly not attempted to rape her. The documents prove that there was a time when she did seem to know what had happened, and a time soon afterward when she became the mouthpiece of a monstrous lie.[15]

Now, half a century later, Carolyn offered up another truth, an unyielding truth about which her tragic counterpart, Mamie, was also adamant: "Nothing that boy did could ever justify what happened to him."

2.

BOOTS ON
THE PORCH

It was probably the gunshot-thud of boots on the porch that pulled Reverend Moses Wright out of a deep sleep about two in the morning on Sunday, August 28, 1955.[1] Wright was a sixty-four-year-old sharecropper, short and wiry with thick hands and a hawksbill nose. An ordained minister in the Church of God in Christ, Wright sometimes preached at the concrete-block church tucked into a cedar thicket just a half mile away; most people called him "Preacher." Twenty-five white-tufted acres of cotton, almost ready for harvest, stretched out behind his unpainted clapboard house in a pitch-black corner of the Mississippi Delta called East Money.[2] He had lived his entire life in the Delta, and he had never had any trouble with white people before.

The old but well-built house would be called a "shack" in a certain stripe of sympathetic news story, but it was the nicest tenant house on the G. C. Frederick Plantation. Mr. Frederick respected Reverend Wright and let his family occupy the low-slung four-bedroom house where he had lived himself before he built the main house. Its tin roof sloped toward the persimmon and cedar trees that lined the dusty road out front. A pleasant screened-in porch ran its entire face. From the porch two front doors opened directly into two front bedrooms; there were two smaller bedrooms stacked behind those.[3]

The accounts of what happened in the Wright home that morning vary slightly, but the interviews given to reporters soon after the event seem to be the most reliable. "Preacher! Preacher!" someone bellowed from inside the screened porch. It was a white man's voice. Wright sat up in bed. "This is Mr. Bryant," said another white man. "We want to talk to the boy. We're here to talk to you about that boy from Chicago, the one that done the talking up at Money."[4] Wright thought about grabbing his shotgun from the closet; instead he pulled on his overalls and work boots and prepared to step outside.[5]

Still asleep were his three sons, Simeon, Robert, and Maurice; his wife, Elizabeth; and three boys from Chicago visiting for the summer: his two grandsons, Curtis Jones and Wheeler Parker Jr.; and his nephew Emmett, whom they all called "Bobo." Somehow Wright had gotten wind of a story involving Bobo at Bryant's Grocery and Meat Market in Money. At first Wright had feared trouble might come of it, but the vague details seemed trifling and convinced him that repercussions were unlikely.[6] Otherwise he would have put his niece's boy on the next train home. Now that he had angry white men at his door, he decided to stall, hoping that Bobo would scamper out the back door and hide. Then Wright would tell the men that the boy had taken the train for Chicago on Saturday morning. "Who is it?" he called out.[7]

In the darkness Wright heard rather than saw Elizabeth head quickly for the two back rooms to wake the boys. Simeon slept in one of the blue metal beds with his beloved cousin Bobo.[8] Robert slept in another bed in the same room. Curtis stayed by himself in the other back room. In the second front bedroom the two sixteen-year-olds, Wheeler and Maurice, shared a bed. Eight people in mortal danger.[9]

Elizabeth later told reporters, "We knew they were out to mob the boy." There was neither time nor necessity to talk about what to do. Her only recourse was obvious: "When I heard the men at the door, I ran to Emmett's room and tried to wake him so I could get him out the back door and into the cotton fields."

Wright slowly stepped out of his bedroom and onto the porch, closing the door behind him. In front of him stood a white man, six feet two inches and weighing 250 pounds. "That man was Milam," the minister said later. "I could see his bald head. I would know him again anywhere. I would know him if I met him in Texas."[10] In his left hand the imposing Milam carried a heavy five-cell flashlight. He hefted a U.S. Army .45 automatic in his right.[11]

Wright did not recognize the rugged-looking man, six feet tall and perhaps 190 pounds, who had identified himself as "Mr. Bryant" and stood just behind Milam, though his small grocery store was not three miles distant.[12] Wright could see that he, too, carried a U.S. Army .45. When both men pushed past him into the house, he could smell them; at that point they had been drinking for hours.[13]

Standing by the door just inside the screened porch, a third man turned his head to one side and down low, "like he didn't want me to see him, and I didn't see him to recognize him," the preacher said.[14] Wright assumed the third man was black because he stayed in the shadows, silent: "He acted like a colored man."[15] This was likely one of the black men who worked for Milam. Or, if Wright's intuition was mistaken, it might have been a family friend of the Milam-Bryant family, Elmer Kimbell or Hubert Clark, or their brother-in-law Melvin Campbell.[16]

Echoing Bryant, Milam said, "We want to see the boy from Chicago."[17]

Wright slowly and deliberately opened the other bedroom door, the one leading into the front guest room where the two sixteen-year-olds slept. The small room quickly became crowded and thick with the odors of whiskey and sweat; faces, guns, and furnishings were caught in the shaky and sparse illumination of Milam's flashlight. "The house was as dark as a thousand midnights," Wheeler Parker recalled. "You couldn't see. It was like a nightmare. I mean—I mean someone come stand over you with a pistol in one hand and a flashlight, and you're sixteen years old, it's a terrifying experience."[18]

Milam and Bryant told Wright to turn on some lights, but Wright only mumbled something about the lights being broken.[19] The wash of the flashlight swept from Maurice to Wheeler and back to Wright. The white men moved on. "They asked where the boy from Chicago was," recalled Maurice.[20]

"We marched around through two rooms," Wright recounted. Milam and Bryant, clearly impatient, may have suspected Wright was stalling. Elizabeth had moved quickly to wake Emmett, but he moved far too slowly. "They were already in the front door before I could shake him awake," she said.[21]

Now the two white men stood over the blue metal bedstead where the fourteen-year-old boy from Chicago lay with his cousin. "Are you the one who did the smart talking up at Money?" Milam demanded.

"Yeah," said Emmett.

"Well, that was my sister-in-law and I won't stand for it. And don't say 'Yeah' to me or I'll blow your head off. Get your clothes on." Milam told Simeon to close his eyes and go back to sleep, while Emmett pulled on a white T-shirt, charcoal gray pants, and black loafers.[22]

Elizabeth offered them money if they would leave the boy alone. Curtis thought Bryant might have accepted if he had been there without his burly half-brother, but Milam yelled, "Woman, you get back in the bed, and I want to hear them springs squeak." With unimaginable poise Wright quietly explained that the boy had suffered from polio as a child and had never been quite right. He meant no harm, but he just didn't have good sense. "Why not give the boy a good whipping and leave it at that? He's only fourteen and he's from up North."[23]

Milam turned to Wright and asked, "How old are you, Preacher?"

Wright answered that he was sixty-four. "You make any trouble," said Milam, "and you'll never live to be sixty-five."[24]

Milam and Bryant hauled the sleepy child out the front door toward a vehicle waiting beyond the trees in the moonless Mississippi night. Wright could hear the doors being opened, though no interior light came on; then he thought he heard a voice ask "Is this the boy?" and another voice answer "Yes." He and others later speculated that Carolyn Bryant had been in the vehicle and had identified Emmett, thereby becoming an accessory to murder. But besides being dark it was hard to hear the low voices through the trees, and Wright told reporters at the time, "I don't know if it was a lady's voice or not." The vehicle pulled away without its headlights on, and nobody in the house could tell whether it was a truck or a sedan.

After he heard the tires crackling through the gravel, Wright stepped out into the yard alone and stared toward Money for a long time.[25]

3

GROWING UP BLACK IN CHICAGO

It was Reverend Wright who started the three Chicago boys, Emmett, Curtis, and Wheeler, thinking about going to Mississippi that summer of 1955, only a few days after Emmett turned fourteen. A former parishioner, Robert Jones, who was the father-in-law of Wright's daughter, Willie Mae, had passed away in Chicago, and the family asked Wright to conduct the funeral. While he was up north it was decided that he would bring Wheeler and Emmett back to Mississippi with him and that Curtis would join them soon afterward.[1]

The image of Wright in Chicago is one of the more pleasing in this hard story. While he was in town he rode the elevated train, toured the enormous Merchandise Mart and the downtown Loop, and gazed out from atop the 462-foot Tribune Tower, which featured stones from the Great Pyramid, the Alamo, and the Great Wall of China, among other famous constructions. He enjoyed the sights but was hardly dumbstruck.

The city had its glories, he acknowledged, but he boasted of the simple pleasures of rural life in the Delta. Four rivers—the Yazoo, the Sunflower, the Yalobusha, and the Tallahatchie—passed near his Mississippi home, and there were seven deep lakes. This surely offered the best fishing in the world.[2] His stories enchanted Emmett. "For a free-spirited boy who lived to be outdoors," Emmett's mother, Mamie, said, "there was so much possibility, so much adventure in the Mississippi his great-uncle described." Although Mamie originally refused to let him go south, she soon relented under a barrage of pressure from Emmett, who recruited support from the extended family.[3]

One stock theme in stories of Emmett Till is that, being from the North, he died in Mississippi because he just didn't know any better. How was a boy from Chicago supposed to know anything about segregation or the battle lines laid down by white supremacy? It is tempting to paint him, as his mother did, as innocent of the perilous boundaries of race; her reasons for doing so made sense at the time, even though being fourteen and abducted at gunpoint by adults would seem evidence enough of his innocence. But it defies the imagination that a fourteen-year-old from 1950s Chicago could really be ignorant of the consequences of the color of his skin.

Race worked in different ways in Chicago than it did in Mississippi, but there were similarities. After Emmett was murdered one newspaper writer, Carl Hirsch, had the clarity of mind to note, "The Negro children who live here on Chicago's South Side or any Northern ghetto are no strangers to the Jim Crow and the racist violence. . . . Twenty minutes from the Till home is Trumbull Park Homes, where for two years a racist mob has besieged 29 Negro families in a government housing project." Emmett attended a segregated, all-black school in a community "padlocked as a ghetto by white supremacy." Hirsch pointed out, "People everywhere are joining to fight because of the way Emmett Till died—but also because of the way he was forced to live."[4]

There was at least one way that Chicago was actually more segregated than Mississippi. A demographic map of the city in 1950 shows twenty-one distinct ethnic neighborhoods: German, Irish, Swedish, Norwegian, Dutch, Czech and Slovak, Scottish, Polish, Chinese, Greek, Yugoslavian, Russian, Mexican, French, and Hungarian, among others.[5] These ethnic groups divided Chicago according to an unwritten treaty, which clearly stated that Germans, for instance, would live on the North Side, Irish on the South Side, Jews on the West Side, Bohemians and Poles on the Near Southwest Side and Near Northwest Side, and African Americans in the South Side's "Black Belt." All of these groups had gangs that regarded their neighborhood as a place to be defended against encroachments by outsiders. And the most visible outsiders were African Americans.

Black youngsters who walked through neighborhoods other than their own did so at their peril. Those searching for places to play, in parks and other public facilities, were especially vulnerable. These were lessons that black children growing up on the South Side learned with their ABCs.[6]

Like many of his contemporaries, Emmett loved baseball. "He was a nice guy," said thirteen-year-old Leroy Abbott, a teammate on the Junior Rockets, their neighborhood baseball team. "And a good pitcher—a lot of stuff on the ball."[7] With the White Sox and the Cubs both in Chicago, it may seem odd that Emmett rooted for the Brooklyn Dodgers, but for a young black baseball enthusiast they were hard to resist. Brooklyn had not only broken the color barrier by signing Jackie Robinson in 1947 but had also signed the catcher Roy Campanella the next year and in 1949 acquired Don Newcombe, Emmett's hero. Newcombe soon became the first black pitcher to start a World Series game and the first to win twenty games in a season.[8]

One night when Emmett was about twelve, Mamie sent him to the store to buy a loaf of bread. He was ordinarily reliable about such things, but on the way home he saw some boys playing baseball in the park. He walked over to the backstop and talked his way into the game. He planned to stay for a short time and then go home with the bread; his mother might not even notice, he told himself. But his passion for the game overcame him; he must have become absorbed in the smell of the grass and the crack of the bat, the solid slap of the ball into leather and the powdery dust of the base paths. "So, I guess he just put down the bread and got in that game," his mother recalled. "And that's exactly where I found Bo—Bo and that loaf of bread. Of course, by that time, the bread kind of looked like the kids had been using it for second base."[9]

Emmett was a lovable, playful, and somewhat mischievous child but essentially well-behaved. He spent his early years in Argo, less than an hour's train ride from his eventual home in Chicago, and was unusually close to his mother and other family members. But he grew up in one of the toughest and most segregated cities in America, knowing as virtually every African American in Chicago knew that in Trumbull Park black fathers kept loaded firearms in their home for good reason. Emmett did not have to go to Mississippi to learn that white folks could take offense even at the presence of a black child, let alone one who violated local customs.

The *City of New Orleans* was the southbound train of the Illinois Central Railway that would carry Emmett to Mississippi in August 1955. The Illinois Central connected Chicago to Mississippi not merely by its daily arrivals and departures but also by tragedy, hope, and the steel rails of history. Over the six decades from 1910 to 1970 some six million black Southerners departed Dixie for promised lands all over America. Chicago, the poet Carl Sandburg wrote, became a "receiving station and port of refuge" for more than half a million of them, vast numbers of whom hailed from Mississippi. "The world of Mississippi and the world of Chicago were intertwined and interdependent," writes the historian Isabel Wilkerson, "and what happened in one did not easily escape notice of the other from afar." Straight up the line of the northbound Illinois Central the carloads of pilgrims from the Delta would rumble, the floors littered with so many empty pasteboard boxes that had been lovingly packed with food from back home that people called it "the chicken bone express." These migrants brought with them musical, culinary, religious, and community traditions that became a part of Chicago; in fact, the narrow isthmus on the South Side where African Americans were confined was often referred to as "North Mississippi."[10] What they found there, however, was not the Promised Land. Though Chicago offered a welcome breath of free air, the newcomers also faced a relentless battle with the white working class over neighborhood borders and public space.

The first wave of the Great Migration, from 1910 to 1930, doubled the number of African Americans in Chicago, placing them in competition for jobs and space with earlier generations of migrants, most of them from central and southern Europe. Herded into the South Side, quickly overwhelming its capacity, the descendants of enslaved Southerners overflowed the ghetto's narrow confines. Housing shortages pushed them over invisible racial boundaries into formerly all-white neighborhoods, where they confronted threats and violence. One 1919 study of race relations in Chicago called these upheavals "a kind of guerilla warfare." Between July 1917 and March 1921 authorities recorded fifty-eight bombings of buildings bought or rented by African Americans in formerly all-white sections of the city.[11]

On Sunday, July 27, 1919, a black seventeen-year-old named Eugene Williams drifted across one of those invisible boundaries and set off a small race war. As he and his friends swam at a segregated beach on Lake Michigan, their wooden raft floated into "white" water. A white man threw rocks at them, hitting Williams in the head, causing him to sink and drown. Rather than arrest the assailant, white police officers hauled off a black bystander who objected to their inaction. Soon carloads of white gunmen raced through the African American neighborhoods, spraying bullets. Black snipers returned fire. Mobs of both races roamed the streets, stoning, beating, and stabbing their victims. The riot raged for five days in that notorious Red Summer of 1919; police shot down seven African Americans, white mobs killed sixteen more, and black mobs killed fifteen whites. Thousands became homeless as a result of arson, and more than five hundred citizens, two-thirds of them black, were seriously injured.[12]

The politics of "the New Negro" were in evidence even before the upheavals but were far more prominent in Chicago afterward, in a direct response to the race riots.[13] Though mourning the deaths, African

Americans in Chicago were proud that they had risen up to defend their lives and communities. Added to that, pride in the patriotic sacrifices and military achievements of black soldiers in World War I met a new determination to make America itself safe for democracy.[14] W. E. B. Du Bois, who had urged African Americans at the outset of the war to lay aside their special grievances and support the war effort wholeheartedly, wrote:

> *We return.*
> *We return from fighting.*
> *We return fighting.*
> *Make way for Democracy! We saved it in France, and*
> *by the Great Jehovah, we will save it in the United*
> *States of America, or know the reason why.*[15]

Du Bois's *Crisis* magazine, which had a circulation of 385,000 in 1915, sold 560,000 copies in the first six months of 1917.[16] Marcus Garvey's Universal Negro Improvement Association had awakened the spirit of black pride and self-assertion on a scale unprecedented, and the charismatic Jamaican black nationalist's movement swelled across the country, including a flourishing UNIA chapter in Chicago.[17] African American parents began to buy dark-skinned dolls for their children and to sing what in 1919 became known as the "Negro National Anthem," penned years earlier by the NAACP's James Weldon Johnson:

> *Lift every voice and sing*
> *Till earth and heaven ring*
> *Ring with the harmonies of liberty. . . .*
> *Facing the rising sun of our new day begun*
> *Let us march on until victory is won.*

The circulation of "race" publications skyrocketed.[18] The *Chicago Defender*'s rose from 10,000 to 93,000 in the war years alone, making it the largest-circulation black newspaper in America. The *Defender* shipped two-thirds of its issues outside Chicago, most of them to Mississippi.[19] "On our porches we read the *Chicago Defender*," recalled Mississippian Helen O'Neal-McCrary, "the only news that black people in Clarksdale could read and believe."[20]

"I did not understand the restrictive soreness imposed by segregation," wrote a summertime visitor from Mississippi, "until I got off that train and breathed the freer air of Chicago."[21] If he had stayed longer, however, this temporary migrant might have grown disillusioned. In the decades after the bloody conflict of 1919, the color line in Chicago was even more sharply drawn. The South Side became almost totally black and the North Side almost entirely white. The Chicago where Emmett Till grew up became one of the most racially divided of all American cities and would remain so into the twenty-first century.[22]

By the 1940s Chicago led the nation in the use of racial covenants on real estate; these restrictions on who could buy property and where they could buy it covered roughly half of the city's neighborhoods. Realtors generally refused to show homes to buyers except in neighborhoods occupied by people of their own race. Many African Americans, regardless of their means, could not get a mortgage and became ensnared in a vicious contract-based buying system that routinely ended up bankrupting them. Federal Housing Authority mortgage insurance policies strengthened Chicago's racial boundaries by denying mortgage insurance and home improvement loans to any home on a "white" street after even one black family moved in. Later the Chicago activist Saul Alinsky sardonically defined *integration* as "the period of time between the arrival of the first black and the departure of the last white."[23]

Various "neighborhood improvement associations" and street gangs fought to keep their neighborhoods all-white; racially motivated residential bombings were one preferred method in the late 1940s and early 1950s. In 1949 a mob of two thousand whites attacked a small apartment building in Park Manor, a white neighborhood on the South Side, after a black couple had purchased the building. Violence flared again in 1951, when five thousand whites spent several days firebombing and looting a building in suburban Cicero after the owners rented a single unit to a black family. The governor of Illinois dispatched the Illinois National Guard to quell the riot, which injured nineteen people. In 1954 the Chicago Housing Authority acknowledged that "bombings are a nightly occurrence" where African American families had moved into neighborhoods that white people regarded as their own.[24]

In 1948 the Chicago Urban League reported that 375,000 black residents of the South Side lived in an area that could legally accommodate 110,000. The overpopulation led to abysmal sanitary and health conditions, and many of the buildings were firetraps. Overcrowding pushed hard against the racial boundaries that encircled the African American areas; between 1946 and 1953 six episodes of riots involving between one thousand and ten thousand people followed efforts of black citizens to move into areas such as Cicero, Englewood, and Park Manor.[25] In neighborhood after neighborhood a familiar drama played out along the hard lines of Chicago segregation. An aspiring black family seeking to escape the ghetto agreed to pay an inflated price for a home on a previously all-white block. Alarmed white residents would quickly sell their homes, allowing landlords to gobble them up at bargain prices. The landlords would then subdivide the apartments and houses into kitchenettes and rent them to blacks, substantially increasing the combined rental income for the building. Neglecting repairs and maintenance, the landlords—the entire local real estate industry, really—created the same ghetto conditions that the African American pioneers had fled at the start in the first act of this three-act tragedy. "In Chicago's 'bungalow belt,' where a large number of European ethnic working-class families owned their own homes," observes the historian and cultural critic Craig Werner, "the first signs of the depressing pattern understandably generated fierce resistance. The result was what one historian called 'chronic urban guerilla warfare.'"[26]

The worst and longest-running of the Chicago housing conflicts lasted from August 4, 1953, until well into the fall of 1955.[27] It began when Donald and Betty Howard and their two children moved into Trumbull Park Homes, a 462-unit development in South Deering near the steel mills. The project had been kept all-white since opening in 1939; the light-skinned Betty Howard got in because the Chicago Housing Authority misidentified her during the requisite interview. By August 9 a mob of two thousand angry whites was throwing bricks and firebombs and Donald Howard was guarding his family's apartment with a rifle. The white vigilantes used fireworks to harass and intimidate the Howards at night. Though police cars shuttled the family in and out of Trumbull Park, once the Howards were in their home the officers did little to ensure their safety, simply yielding the streets to the mob. On August 10 the white mob stoned thirty passing black motorists and attacked a city bus carrying African Americans, nearly tipping it over before police intervened. Throughout it all Mayor Martin Kennelly said nothing about the ongoing violence.

As the number of African American families moving into Trumbull Park increased to ten, the so-called South Deering Improvement Association kept the riots rolling and organized economic reprisals against any neighborhood stores that served African American customers. The city parks became particular battlegrounds; when black youths tried to use a baseball diamond in the neighborhood, the Chicago Police Department had to dispatch four hundred officers to protect them. In protest Willoughby Abner, a trade unionist who was president of the Chicago NAACP, organized a baseball "play-in" at South Deering's main park; the United Packinghouse Workers of America, an interracial but increasingly black union devoted to civil rights, provided support as Abner mobilized the NAACP and sued the city for inaction. Still conditions in South Deering did not change appreciably; in late 1954 Chicago's Federal Housing Administration director called Trumbull Park "a running sore in our civic life."[28]

White residents believed that this black "incursion" was only the opening gambit of a campaign of racial infiltration: soon African Americans would buy private homes, causing property values to plummet, and start taking "white jobs" in the Wisconsin Steel Works nearby. The heart of the violent white response, however, was more visceral: like many whites in the Deep South, South Deering's white residents had a horror of interracial sex. The *South Deering Bulletin* declared, "White people built this area [and] we don't want no part of this race mixing." A housing inspector sent to South Deering reported that white residents insisted, "It won't be long now and Negroes and whites intermarrying will be a common thing and the white race will go downhill." The South Deering Improvement Association openly rallied whites for the ongoing riots at Trumbull Park by promoting "this fight against forced integration and mongrelization."[29] Walter White of the NAACP saw the sad irony: although African Americans fled Mississippi to escape from racial terror, the violence in Chicago revealed that "Mississippi and the South [followed] them here."[30]

But if the battle against integration in Chicago took on some of the same themes as the battle in Mississippi, there was one big difference: African Americans in Chicago could vote. So when Mayor Kennelly ignored complaints about mob violence and housing segregation and forgot that African Americans had considerable force in the Democratic Party, it cost him his job. Kennelly ran afoul of U.S. Representative William Dawson, the most powerful black elected official in America, who headed the black political machine that remained a crucial part of the larger Chicago Democratic machine. Dawson supervised scores of African American ward committee members, precinct captains, and election workers. He swapped black ballots for patronage jobs and for protection of the lucrative South Side numbers rackets and jitney cabs, a major source of his political funds. When Kennelly's police department targeted the numbers games and jitneys, Dawson declared war. His opposition to Kennelly permitted another Irishman, Richard Daley, head of the Democratic machine, to slide into place.[31]

Chicago insiders expected that African Americans would see major changes along the color line if Daley were elected mayor. So some black voters must have been taken aback when it became clear that Daley's vision for Chicago rested on his commitment to racial segregation in schools and housing. Others may have been disappointed to discover that Representative Dawson shared that commitment, though for different reasons.[32]

For Dawson it was simple enough: he did not want to disperse the black voters whose ballots were the source of his power. Packed into the South Side's State Street Corridor, black voters were manageable. Likewise these ghettoes were where people played the numbers and where the lack of public transportation made unlicensed jitneys an essential part of life; both of these illicit operations poured money into Dawson's campaign coffers. In exchange for his ability to deliver black votes, Dawson expected that Daley would keep the police away from the numbers runners and jitney drivers. He also expected Daley to allot him a share of the city's patronage jobs. Thus, as far as Dawson was concerned, the preservation of the racial status quo was a practical necessity and good business.[33]

Daley, on the other hand, reflected the stony conservatism that prevailed in most white, ethnic, working-class neighborhoods in the 1950s. He believed in racial separation of the kind that marked his own Irish neighborhood of Bridgeport and the various ethnic neighborhoods that bordered it, especially the South Side's black ghetto. Black people belonged on the other side of Wentworth Avenue, and that was that. He came to power at a time when the black population hit record highs, when Chicago's white middle class and a good many downtown businesses had begun to flee to the suburbs, aided by cheap FHA loans, lower taxes, and America's new highways. "White" neighborhoods became "black" neighborhoods as poor African Americans flooded in from the South. Daley intended to rescue Chicago from this dynamic by building a new city on an unarticulated commitment to segregation.

However, the African American vote marshaled by Dawson's machine was too rich for a machine Democrat like Daley to ignore, so he carefully appealed to both sides on the dicey issues around race. He was solicitous of Dawson and made it clear that the numbers rackets and the unlicensed jitneys would encounter no legal hassles under a Daley administration. He presented himself as a civil rights supporter in the black community, even giving lip service to the notion that everyone had a right to live wherever their talents would take them.

Through the white grapevine, however, he spread the word that he would preserve the color line in housing. He made quiet racial appeals in the white working-class neighborhoods, circulating letters from the nonexistent "American Negro Civic Association" that praised his opponent for supporting open housing. He spoke in favor of public housing but always added, "Let's not be arguing about where it's located." He appointed a committee to study the racial problems at Trumbull Park Homes but made sure the group did nothing.

Daley rode into office on a heavy majority of black votes on April 20, 1955, four months to the day before Emmett Till climbed onto the *City of New Orleans* with his great-uncle Moses Wright and his cousin Wheeler Parker and set out for Mississippi. In the fall, well after the election, the NAACP's Willoughby Abner brought five thousand demonstrators to city hall holding signs that protested the city's ongoing racial segregation: "Trumbull Park—Chicago's Little Mississippi."[34]

4

EMMETT IN CHICAGO AND "LITTLE MISSISSIPPI"

Mamie Carthan, a bright, plump toddler, was born in Webb, Mississippi, "really not much of a town at all," she remembered, more like a handful of stores "in search of a town." The main street divided the black and white sides of the dusty little community. "Just about any place else would have been better than Mississippi in the 1920s," she mused. In 1924 the Great Migration swept Alma and Wiley Nash Carthan and their two-year-old daughter to Argo, Illinois, a town of fewer than three thousand people some twelve miles from Chicago. Wiley had landed a job at Argo's central enterprise, the Corn Products Refining Company.

It wasn't long before the Carthans referred to Argo as their own "Little Mississippi." Other family members had already established a beachhead for relatives, friends, and even strangers who heard there might be work in Chicago or Argo. Mamie's grandmother founded a church for the migrants. "As I was growing up," Mamie wrote later, "it really seemed like almost everybody from Mississippi was coming through our house—the Ellis Island of Chicago."[1]

Mississippi in memory remained both the ancestral homeplace and a land of ghosts and terror. "All kinds of stories came out of Mississippi with the black people who were running for their lives," Mamie wrote. There had been talk of a terrible lynching in Greenwood, another young man strung mutilated from a tree, not far from Money, where her uncle Moses and aunt Elizabeth Wright lived. The Greenwood lynching "was the sort of horrible thing you only heard about in the areas nearby." In the decades before the civil rights era, racial killings in remote corners of the Deep South frequently went unreported by the national or even the local press.[2] What the migrants learned by word of mouth has since been established as fact. Mississippi outstripped the rest of the nation in virtually every measure of lynching: the greatest number of lynchings, the most lynchings per capita, the most lynchings without an arrest or conviction, the most female victims, the most multiple lynchings, and on and on.[3] Richard Wright, writing of his boyhood in Mississippi in the 1920s, observed, "The things that influenced my conduct as a Negro did not have to happen to me directly; I needed but to hear about them to feel their full effects in the deepest layers of my consciousness. Indeed, the white brutality that I had not seen was a more effective control of my behavior than that which I knew."[4]

Mamie was haunted by the story of a little black girl who had been playing with a white girl at the home of the white family that employed her mother. The white girl got upset with the black girl and ran to tell her father as he walked up the driveway from work. He angrily snatched up the black girl, shook her like a rag doll, then tossed her up against a tree in the front yard. "Now, that girl's mother had to finish her day's work before she could even look after her daughter, who was left there writhing in pain the rest of the day," Mamie remembered many years later. "Eventually, the little girl died of her injuries." This was "a cautionary tale," she said, a tale of horror rooted in real experience, whether or not it was precisely true in its particular details. "Was this a true story? I don't know. But I do know this: Somewhere between the fact we know and the anxiety we feel is the reality we live."[5]

Though Argo was almost close enough to be a part of Chicago, the reality the Carthan family lived was "a sleepy little town where whites called blacks by their first names and where blacks would never dare to do the same thing."[6] Segregation was haphazard and unpredictable. "Some of the white kids and people of Argo could get nasty and give the Negro children a rough time just for spite," recalls Gerald V. Stokes, who grew up in Argo during the 1950s. Black children in Argo were told never to enter restaurants or business establishments except in the company of an adult: "They were warned never to take short cuts to school through the white neighborhoods. They were warned never to talk to strangers, especially white strangers, or to talk back [to] white folks." In Argo, according to Stokes, "bad things happened to little Negro children at the hands of white strangers."[7]

Argo was a community of immigrants, nearly all of them from Mississippi, who had come in search of the Promised Land and found something less grand. Even so the kids frolicked loudly up and down the streets until darkness began to fall. Mothers and fathers sat on the stoops and laughed out loud without concern for who might hear. Neighbors shared access to telephones, which were rare, and simply yelled to communicate with friends across the street. "Everyone talked loudly and freely," writes Stokes. "They felt secure in the blackness of their lives."[8] Mamie agrees: "You could pretty much see it all from our end, on the sidewalk in front of our home. The elementary school right across the street, the church not too far down the street, and to the right, filling up the distant horizon, was the Corn Products plant. Our whole world."[9]

All the Carthans' immediate neighbors were family members from Mississippi. Aunt Marie, Uncle Kid, and his cousin "June Bug" lived west of the Carthans. Uncle Crosby and his family lived to the east. Just behind them were Aunt Babe and Uncle Emmett. Mamie's great-uncle Lee

Green lived across the street. This must have made a terribly hard thing a little easier when Mamie's father left the family in 1932, when she was eleven, and moved to Detroit to marry another woman. Even with her daddy gone, Mamie grew up in the bosom of this extended family, much beloved, secure from the many-sided perils of Mississippi and sheltered from the Chicago that E. Franklin Frazier called "the City of Destruction."[10]

And it wasn't just Mississippi kin who found welcome and a sense of belonging in the Carthan household. "Our house was the meetinghouse, the gathering place, the center of the community," recalled Mamie, whose deep attachment to her mother never faded. "It was the place where Mama had helped [her mother] found the Argo Temple Church of God in Christ and where she recruited new members with practically each new Mississippi migrant." Here in "Little Mississippi" they found new lives anchored by the Corn Products plant and surrounded by the family but still tethered to the South.[11] In the summer black folks from Argo felt an unwritten obligation to visit family in Mississippi and reconnect with the social and spiritual world that continued to define their lives.[12]

On October 14, 1940, at eighteen, Mamie married Louis Till, a burly, athletic gambler who favored dice and poker and loved boxing. "I became pregnant right away, and being the plump type, I began filling out rapidly," she later told a newspaper reporter. "This set the neighbors' tongues to clacking busily." Emmett Louis Till was born in Cook County Hospital on July 25, 1941, after a long and difficult labor. Medical complications during a breech birth left the infant scarred from the instruments and with a badly bruised knee, among other injuries. The doctors thought that some damage might be permanent; thankfully they were wrong. Mamie was in bad shape for months, but at two months old Emmett "was a beautiful baby with a sunny disposition and every sign of being normal." By Mamie's account, her husband never once came to see his son or his wife in the hospital.[13]

The family had nicknamed the baby "Bobo" even before he was born, and the name stuck. Because he was born so light-skinned, with blond hair and blue eyes, the neighbors gossiped; the milkman and the ice man both became suspects, according to Mamie. Soon, however, Emmett's hair darkened, his eyes turned hazel brown, and he favored Louis so strongly that his paternity could not be doubted.[14]

Louis Till, dictatorial and ill-tempered, resented the amount of time his wife spent at her mother's house after Emmett was born. He expected to have supper waiting for him when he came home from his Corn Products job. If Mamie and the baby were still at her mother's, he became abusive and violent when she did return. During one episode Mamie threw boiling water on him. Eventually she and the baby moved into her mother's house, and in 1942 Mamie and Louis separated permanently.[15]

Louis's efforts to reconcile with Mamie also became abusive, and she obtained a restraining order against him, which he violated persistently. In later years she claimed that a judge finally gave him a choice between jail and military service. He joined the army and regularly sent his estranged family child support payments of $22 a month. In July 1945 the checks stopped coming and a telegram from the Department of Defense informed her that Louis had been executed in Italy for "willful misconduct." The army later sent her attorney a record of the court-martial, which would have explained that he had been convicted of raping two women and killing a third while stationed in Italy. The army shipped his few belongings to Argo, including a silver ring engraved with the initials "L.T." She put the ring away, thinking that Emmett might want it someday.[16]

When Emmett was naughty, he hid under the bed, peeking out to see if anyone was chasing him. Mamie was completely disarmed by his playful defiance and could hardly think to scold him.[17] Her mother supplied most of the order and discipline in the household. As far as Alma Carthan was concerned, she now had two babies. "I was the big kid, Emmett was the little kid," Mamie explained. "We were so much like brother and sister, like friends back then, and it added a unique dimension to the mother-son bond we would forge over the years ahead."[18]

That bond grew even stronger after a crisis that struck them in the summer of 1946. Emmett had just turned six. His mother noticed that although he had plenty of energy during the day, he seemed to deflate with fatigue every evening. This was unlike him, who was ordinarily a dynamo until bedtime. Then his temperature began to rise sharply every night. Alma and Mamie rubbed him with "goose grease" and made him drink "hoof tea" every evening. "These remedies were supposed to cure a lot of things," Mamie wrote. "I never knew why or how. I didn't even know what kind of hoof came in that little box of tea. I didn't know what [goose grease] was supposed to do, either. I just knew that all our folks from Mississippi used it."[19]

But Emmett only grew worse with the home remedies, so they called a doctor, who gave the boy a diagnosis that broke Mamie's heart: "Polio was the worst thing that could happen to you back then. It didn't kill you, but it could take your life away from you just the same." Polio threatened permanent limb damage and lifelong disability. The doctor ordered Emmett quarantined at home; he couldn't leave the house, nor could anyone come over to see him, a decree the six-year-old fought. "Mama had to sit with Emmett all the time, practically holding him in the bed," remembered Mamie.[20]

After thirty days "he had beaten it. . . . He was finally up and running again and practically tore a hole in the screen to get out."[21] The polio left him with a noticeable stutter and weak ankles that forced him to wear special shoes, but neither of those infirmities kept him from moving relentlessly through the world. Endearing and lively, he had plenty of toys and plenty of friends, so that his grandmother's yard became a kind of neighborhood playground. He and his friends played baseball and basketball in the nearby schoolyard and park, and on special occasions they liked to go to the Brookfield Zoo about three miles away.

Just as Emmett recovered from polio and escaped to the baseball diamond, Mamie's cousins Hallie and Wheeler Parker Sr. moved from Mississippi to the house next door. It was the house where her uncle Crosby Smith had lived until he decided to move back to Mississippi. Emmett became best friends with Wheeler Jr., who was two years older.[22]

Aside from that month of pain and the loneliness of his quarantine, those three years in Argo, from 1947 to 1950, were paradise for Emmett, surrounded by playmates and next door to his best friend. So when Mamie and Emmett moved to Detroit in 1950 to live with her father while she worked at the Fort Wayne Induction Center, Emmett was terribly homesick. The following year she met Pink Bradley, an auto worker at Chrysler, and married him after a brief romance. Seeing Emmett's homesickness grow worse Mamie decided that her boy should move back to Argo and live with Uncle Kid and Aunt Marie.[23] When Pink lost his job at Chrysler, he and Mamie relocated to Chicago; there they moved into an apartment on South St. Lawrence, right next to the apartment her mother had moved into earlier. Emmett happily joined them.[24]

But Pink began spending his weekends in Detroit, a five-hour drive away. Slowly he drifted back to Detroit altogether and away from his marriage. When Mamie learned that he had a woman in Detroit, she and her mother changed the locks and threw his clothes into the yard.[25]

Now Alma, Mamie, and Emmett were back together again, with Emmett, a natural prankster and mimic, keeping all of them entertained. "From the very beginning with Emmett there was laughter," wrote Mamie. "I heard about more chickens crossing more roads, and knock-knock this and knock-knock that. All those tired, old jokes that were still new to him. Sometimes he would tell riddles that he seemed to have been making up, because they didn't make sense. Or maybe you just had to be [a child] to understand them."[26] They got a television, and Emmett learned to mimic the early comedians. "He knew the routines of all the top ones on television," Simeon Wright remembered, "Red Skelton, Jack Benny, Abbott and Costello, and George Gobel. Gobel, with his deadpan delivery, was a particular favorite of Bobo's."[27]

Emmett's audience included the band of boys he ran with on weekends in Argo, which he could reach in less than an hour on the 63rd Street bus. His cousins Wheeler, William, and Milton Parker were regular playmates, as were his cousins Crosby "Sunny" Smith, Sam Lynch, and Tyrone Modiest, and friends like Donny Lee Taylor and later Lindsey

Hill. "When those boys got together it was non-stop laughter," his mother recalled.[28]

Sometimes Mamie would cart the boys to the beach on Lake Michigan; she had to take care, of course, since segregation of public beaches, though not a matter of law, was a fact of life in the 1950s.[29] But that did not spoil their fun. One day the Argo crew decided to wear their swim trunks home and just carry their clothes. "Donny Lee made the mistake of falling asleep in the car with Bo," Mamie wrote. "He woke up to find that he was wearing his underwear after all. On his head."[30]

On warm nights the boys would end up "doo-wopping" under a lamppost. It was most likely Emmett, whom everybody described as a natural showman who "liked the spotlight," who brought this new form of entertainment to his cousins.[31] Throughout the 1950s a burgeoning South Side street-corner doo-wop scene blended gospel harmonies with pop lyrics to produce groups such as the Dells and the Flamingos and singers like Curtis Mayfield and Jerry Butler.[32] According to Mamie, however, the Argo boys were in no danger of hitting the big time: "For this group of boys standing under a curbside spotlight, the music was off-key, it was out of sync, it was perfect. The grace note of their young lives."[33]

Emmett seemed to blossom in the years he lived in Chicago and played in Argo. He enrolled in the fifth grade at James McCosh Elementary School, only two blocks from their place on South St. Lawrence. McCosh was an all-black school with 1,600 students in kindergarten through eighth grade and an interracial faculty. "Emmett was never a discipline problem," the principal told reporters. "He tended to be quiet. As a student, he was average." Teachers noted that he was close to his mother and that he attended church regularly.

The pastor of the Church of God in Christ in Argo—the church founded many years earlier in Alma Carthan's house—observed that young Emmett rode the 63rd Street bus almost every Sunday morning to attend the church where he had grown up. Wheeler Parker Sr., Emmett's uncle and superintendent of the Sunday school, reported that Emmett had a near-perfect attendance record every year.[34] "He liked going to church," said his mother, who attended far less often, "and he was under the influence of his grandmother, a deeply religious woman."[35]

After his mother's marriage to Pink Bradley ended in 1953, Emmett took on more adult household responsibilities. "Since I had to work and make the living," Mamie said, "Bo did all of the housework and laundry. I did the cooking but he even learned to do some of this. He was a good housekeeper."[36] Eva Johnson, who lived next door, recalled "the time he was going to surprise his mother with a cake." It was a yellow cake "made with eggs like pound cake, only using ready-mix." Emmett asked Johnson to come tell him what was wrong with the cake. "Lord 'a' mercy, he'd watched it 'til it began to rise, then he began stirring it! I told him it was ruined [and] he'd better just quit."[37]

His increased responsibilities moved young Emmett quickly to the edge of manhood. "In between taking care of more and more things for me," Mamie noted, "he made sure he took care of his own things." He was meticulous about his clothes. He fancied a straw hat and a tie when he went to church, and even on the ball field he tried to look his best. But if this was to impress girls, he did not appear to be very good at it. At fourteen he had had only one date, and he never really had a girlfriend. He still stammered at times, a lingering effect of polio. He was only five feet four inches tall, but stocky, weighing about 160 pounds. His great-uncle Moses Wright acknowledged, "He looked like a man."[38]

And so it was understandable that his mother gave her assent to the proposed trip to Money but also delivered lengthy lectures about the differences between Mississippi and Chicago. She urged Emmett to avoid conversations with white people, to speak only if spoken to, and to always say "Yes, ma'am" and "Yes, sir" or "No, ma'am" and "No, sir." If a white woman should walk toward you on the sidewalk, take to the street and lower your eyes. Should any dispute arise with any white person whatsoever, humble yourself and agree with them. Emmett protested that he knew all that, that she had already taught him how to act.

In the glare of attention following the murder of her son, Mamie claimed, "This was the first time I had ever really spoken to Emmett about race." Perhaps. "After all, how do you give a crash course in hatred to a boy who has only known love?" Certainly it is true that Emmett grew up blanketed by love, and if he drifted into any of the racial battles around him in Chicago, no record of it exists. But it is unlikely this was the first time she cautioned her son to watch his step. In the segregated and often violent Chicago of the 1950s, Emmett did not need to take a train south to discover that being black made him a potential target, any more than he needed his mother to explain that fact.

Mamie and Emmett were to meet Uncle Moses and Wheeler at Englewood Station about eight o'clock that morning. It was practically around the corner, but even so, the whistle blew while Mamie was buying Emmett's ticket, and they had to run to the train. "He liked to got left," Wright said later. "If he'd taken five minutes later, he'd have missed it." A quick kiss and he was gone, wearing his father's silver ring engraved "L.T." and waving to his mother from the top of the platform.[39]

5

PISTOL-WHIPPING
AT CHRISTMAS

"The color of our skin didn't make any difference when we were young," Carolyn Bryant lied wistfully, looking back over eight decades. In this cheerful falsehood she had a lot of company. But unlike many other white Americans, Carolyn was also capable of honesty and even clarity on matters of race. "As I grew older," she continued, "I learned that it was not okay to have black friends, [though] our parents taught us that everyone deserves respect."[1]

Carolyn Holloway was born on July 23, 1934, a hot, muggy afternoon, on the Archer Plantation, about ten miles down a blistering Delta blacktop from Cruger, Mississippi. A premature baby weighing only four and a half pounds, she was born on her father's birthday.[2]

"We lived in a big house on the plantation," Carolyn told me. "Not *the* big house, just a large house that was built for the plantation manager." Her father, Tom Holloway, had a reputation as an efficient plantation manager and a good cotton farmer, and plantation owners competed for his services, so the Holloway family moved frequently. Moving had become more and more common in the rural South during the Depression, among Southern sharecroppers moving just down the road as well as poor whites taking the "hillbilly highway" to Detroit and African Americans hopping the "chicken bone express" to Chicago. All trekked in search of greater opportunity.[3] Home was a fragile concept for the Holloways, but they never traveled very far. Though they lived on several different plantations in the Mississippi Delta, Carolyn's heart always belonged in Cruger, where her maternal grandparents and Aunt Mabel, her mother's oldest sister, lived.

Besides being a plantation manager Tom Holloway worked as a prison guard in Lambert at one of the outlying camps of the notorious Mississippi State Penitentiary, better known as Parchman Farm. The prisoners "were all black, they were all black men. And my daddy worked at Camp A, the first one. They had a sergeant and a first rider and a second rider. My daddy was the first rider." All the riders were white men, each with a whip, a shotgun, and a rifle, who oversaw the prisoners from horseback. The inmates in their black-and-white-striped uniforms addressed the drivers as "Cap'n."[4]

Camp A had been part of the O'Keefe Plantation in Quitman County until the Mississippi Department of Corrections purchased the land in 1916. In their physical structure, the camps resembled a slavery-era plantation. Black convicts took the place of the enslaved, but otherwise the systems were identical—the same crops grown in the same way under the same discipline. In both instances the farming operation depended on a handful of poor white men to supervise the captive laborers.

The instrument of authority at Mississippi's prison camps was a leather strap, three feet long and six inches wide, nicknamed "Black Annie," hanging from each driver's belt. A former inmate remarked, "They beat hell out of you for any reason or no reason. It's the greatest pleasure of their lives." It was not unusual for the drivers to whip a convict for working too slowly or for breaking a shovel. "The driver seemed to be everywhere, 'directing, scolding, encouraging, or whacking across the shoulders with the whip.'" More formal punishment was a whipping in the evening, in front of all the men, with the victim spread-eagle on the floor. "They whupped us with big, wide strops. They didn't whup no clothes. They whupped your naked butt. And they had two men to hold you [or] as many as they need." Routine offenses like fighting, stealing, and showing "disrespect" to a driver earned five to fifteen lashes; attempts to escape brought the punishment of whipping without limit—whippings that were sometimes fatal. "They'd kill um like that."[5]

During the 1930s and well into the 1950s the lash enjoyed widespread public support among whites in Mississippi. Editors, church groups, public officials, sheriffs, and prison authorities all seemed to support whipping as "the perfect instrument of discipline in a prison populated by the wayward children of former slaves," writes the historian David Oshinsky. These prison camps were "a powerful link to the past—a place of racial discipline where blacks in striped clothing worked the cotton fields for the enrichment of others. And it would remain this way for another half-century, until the civil rights movement methodically swept it away."[6]

No one would expect a little girl to be told about or to understand the realities of work and life at one of the prison camps attached to Parchman, particularly if her father worked there. On the other hand, it stretches the imagination to think that the sensibilities of a man chosen as first driver on Parchman Farm were so delicate they prevented him from joining in the brutal rigors of his job. But Carolyn remained firm that her father had never taken part in those routines. "The sergeant told him he would have to take his turn on whipping night," she said. "My daddy refused to do it. And on whipping night he would come home, and he would go into the bedroom and close the door and go to bed."

In her living room awash in sunlight, decades separating her from the last violent years of Jim Crow Mississippi, Carolyn recalled the moment segregation and white supremacy became sharply drawn imperatives: a bike ride long ago that she didn't take with Barnes Freeman.

Barnes was the son of black laborers. His mother, Annie Freeman, cleaned the Holloways' house and kept the household running smoothly, while Isadore, his dark-skinned father, worked as a hostler on one of the plantations that Carolyn's father managed. A white family did not need to be affluent to afford black servants. When I was growing up in the 1960s this was still a ubiquitous pattern in the South: black domestics made and served dinner to white families, never sitting down to share in the meals, of course. The practice gave rise to another inevitable lie white people told themselves: that black employees were "just like family." The insurance policies, family wills, holiday times, and dinner tables, to say nothing of churches, schools, neighborhoods, and public facilities, told a more honest story. "Barnes was our friend, just like family, almost," Carolyn insisted. "You know, he was all in our house and everything, and I didn't think a thing in the world about it. We just really liked him."

Barnes was "more light-skinned" than his father, "kind of chocolate milk color," Carolyn told me. "I think he was probably kind of large for his age." He was four or five years older than Carolyn, but despite their age difference, they often played together. Barnes's play was inventive, "like putting a rope on an old tire and hanging it from the tree, and Barnes would push us on that tire. We'd sit with our legs through it and he'd swing us." With the distance of time she described Barnes as her favorite companion, with the possible exception of Aunt Mabel, who frequently kept Carolyn at the family homeplace in Cruger, within about ten miles of the series of plantation houses where Carolyn's family lived while she was growing up. Cruger was where she usually played with Barnes.

Aunt Mabel was more doting grandmother than aunt, with never a cross word for her favorite niece. Mabel contracted polio as a child and had limped after she recovered; as an adult she fell and shattered her hip and thereafter spent most of her time in a wheelchair. Carolyn's time with her aunt was therefore often sedentary, a blessing on the hottest days of the summer. A screened porch ran the length of the house and looked out on the dirt road that led toward town, a generous term when applied to Cruger. "It was a little bitty town [that] just had a row of stores and that's it," Carolyn said. Her grandfather Lee Pikes owned some land near Cruger, and "a little shack down on the lake he sold fish out of, and he had a grist mill and ground meal, and he bootlegged liquor." In the scorching, steamy summer heat, Carolyn and Aunt Mabel spent most of their time on the porch, shelling peas, snapping string beans, or just sitting. "I almost never wanted to be inside and was content to sit in the swing on the porch, trying not to move, so I might be cooler."

But every now and then Carolyn chafed against the quieter tempo of her aunt's house. One summer afternoon when she was ten or eleven and Barnes was fourteen or fifteen, just the age of Emmett Till when he ran afoul of Mississippi's customs, she was sitting alone on the porch when Barnes rode his bicycle past the house and waved. "Hey, Barnes," Carolyn yelled. "Where you going?"

"I'm going to the store to get something for my mama," he replied.

"I asked, 'Can I go with you?'" Carolyn told me. "He said, 'Yeah, sure. Jump on the back.' He had this little rack on the back like old-timey bicycles. . . . I darted off the screened porch, slamming the door behind me as usual. And I jumped on the back of his bike."

Almost instantly, before Barnes had a chance to push off and begin their ride, Aunt Mabel rolled out of the house onto the porch, pushed open the screened door, and screeched at the top of her lungs, "Get off that bike and get in this house, right now!"

"I was startled to hear her scream, as she had never raised her voice at me before. Immediately I did as I was told, but I was puzzled, as she seemed to be mad at me." When Carolyn asked why she was upset, her aunt replied, "Because you don't need to be riding with boys on your bike, because people will talk about you."

"It didn't dawn on me at the time," Carolyn wrote many years later, "but the real reason she was upset and yelled at me was because Barnes was a black boy. . . . This was Mississippi. At that time it was okay [for small children] to play with black friends at your home [but] it was completely unacceptable for me to be with Barnes and go off to the store on the back of his bike."[7] She told me, "I don't remember being around him much after that. So maybe he and I both got corrected, you know."

Carolyn was fifteen in 1949, in high school near Cruger, and had just won the school's beauty contest when her father suffered a series of strokes.

The first two weakened him greatly and kept him from working. The final stroke, at age sixty-three, killed him as he sat in their living room. She was bereft, as was her mother, who was only forty-six years old.

Carolyn's mother began training as a nurse, and the plantation's owners were kind enough to let the family remain in the house until she earned her nursing degree. Then the family moved to the nearby small town of Indianola, the seat of Sunflower County. There her mother worked long hours at the hospital while they lived in a small apartment across the road. Carolyn was barely over five feet tall, weighed less than a hundred pounds, and had lovely brunette hair and full lips. Her new classmates seemed to agree that she was movie-star material, for she soon won her second high school beauty contest. To help make ends meet she worked behind the sales counter at the Morgan & Lindsay Variety Store and babysat for a number of local families. When her mother worked late Carolyn took care of her younger siblings.[8]

Their distance from what Carolyn saw as the paradise of the plantation Delta seemed to be echoed in her memories of race. Indianola, where the Citizens' Council would one day be founded, was firmly and vehemently segregated. "There were no black children in school with us—this was no different from the Delta, of course—and no black families in our neighborhood. In fact, the only contact I had with any black people was when I was waiting on them at the Morgan and Lindsay."

In town Carolyn learned the rigid folkways of race. Black people were expected to say "Yes, ma'am" and "Yes, sir" when talking with white people, even whites younger than themselves. Blacks were "actin' up" or "weren't 'in their place'" if "they didn't step aside when someone white passed them on the sidewalk. They better not look any white person in the eye, either. That'd get them punched." Carolyn maintained that in that respect Indianola "was certainly different from the plantation I grew up on," but it may simply have been that her awareness of racial arrangements grew as she got older.

Of course racial separation went deeper than public social arrangements. The idyll of Barnes and the Cruger of her youth was over. "We could no longer have black friends when we lived in Indianola," she wrote. "It was something that was never spoken directly to us but something we understood as the way things were."[9] At least in memory the change from the more intimate racial paternalism that represented one kind of life in the rural Delta seemed to Carolyn tied to her family's loss of a father and their fall from a certain kind of grace.

But for a pretty white girl Indianola had its charms. At her new high school Carolyn soon had a boyfriend with a car. One day he drove her out in the country and offered to show her "the hanging tree." He explained that "a long time ago" white men had hanged black men "when they were actin' up and weren't in their place." Carolyn knew how trivial an offense could constitute a violation of racial mores. "I'm not sure how long ago 'a long time ago' was," she said as she told the story. "But I told him, 'Sure, I want to see the tree.'" On a deserted side road he stopped the car in front of a huge old tree. Up in the thick lower limbs Carolyn detected an ancient length of rope snagged in the trunk, leaving only a foot or so of frayed rope hanging free, tied in a noose.

Carolyn's childhood stories are a narrative of class decline. They establish the understanding that she and her family were a rung or two above the family into which she married because they were capable of a paternalistic generosity toward black people in a way her in-laws were not. There are grains of truth here, to be sure, but it is also a self-exculpating story: if she hadn't gotten mixed up with the Milam-Bryant clan, the stories suggest, this ugly Till lynching and its aftermath would never have happened. That is almost certainly true. There was, she declared, a greater degree of gentility to her upbringing, and indeed her character, than the Milams and Bryants possessed. She described herself as an innocent wandering into a place she didn't quite belong.

The matriarch of a headstrong clan, born Eula Lee Morgan, gave birth to eight sons and three daughters by two different men. Five boys carried the Milam family name: Edward, the oldest; Spencer Lamar, whom they called Bud or Buddy; followed by John William (J.W.), Dan, and Leslie. Their father had been killed in a road construction accident when a gravel pit caved in on him and three other members of the crew.[10] Eula Lee then married a cousin of her late husband, Henry Ezra Bryant, whom everybody called "Big Boy," and had six more children: Mary Louise, who married Melvin Campbell; two twin boys, Raymond and Roy; then Aileen, James, and finally Doris, born with severe mental disabilities.

On ordinary mornings Eula Lee would fix breakfast while Big Boy opened their store, which was next door. Later in the morning Big Boy would come back and eat his breakfast while his wife took care of the store. Then they would both work in the store all day. One morning Eula Lee had eaten her eggs and sausage and waited for her husband, but he did not come home. When she walked over to the store to find him, she found instead an empty cash register, and one of their cars was missing.

"So she called the bank to see about the bank account," Carolyn told me, "and it's been closed down, and he's gone off with another woman." Big Boy and the woman fled to Arkansas, where they lived for seven years, a separation period after which Mississippi law at the time granted an automatic divorce. After he married the woman, they moved back to Mississippi and ran a store in the small community of Curtis Station, forty miles northeast of Eula Lee. Occasionally Roy, who was sixteen or seventeen when his father deserted the family, would drive up there with Carolyn to see him but he was careful to keep these trips a secret, especially from his mother. She swore that if she ever saw "Big Boy" again she would kill him, which was one reason that she carried a .38-caliber revolver everywhere she went. Mrs. Bryant would frequently say, "If I ever see him again," and shake her head, reaffirming her homicidal vow. "The pistol was in her purse," said Carolyn. "Always."

Carolyn first met Roy Bryant at a party when she was fourteen and visiting her oldest sister and her family in Tutwiler. "He was about seventeen and so handsome," she recalled. A few days later Roy dropped by her sister's house and asked her to accompany him and some friends to another party. Carolyn's sister said no, but Carolyn's eyes said yes, which he noted.

She didn't see Roy again until after her father died. "Roy's family had moved to Indianola about the same time that we moved there," she told me with a glimmer of excitement still flickering sixty-five years later. "I was walking home from school one day, and Roy Bryant rolled up beside me in his Forty-nine Chevrolet." He smiled and offered her a ride. "I didn't hesitate one second." Thereafter Roy appeared quite regularly to give her a ride and sometimes take her for a hamburger and a Coke on the way. "I did slip away with him a few times, but I knew I had to get home to babysit."

At eighteen Roy joined the 82nd Airborne and was stationed at Fort Bragg, North Carolina. Juggling school, her job, and her work about the house, Carolyn waited eagerly for his furloughs. "We seemed to grow closer and closer, and I looked forward to his visits when he received leave from the service. I just knew I was in love." One beautiful day in the spring of 1951, Roy proposed. "We decided to elope the next day," she said, because she was only sixteen and hadn't finished high school. "Mama would never sign for me to get married." The next day Carolyn pretended she was going to school but instead met her beau at the post office in Indianola. They picked up a license at city hall, drove to the parsonage at the Second Baptist Church in Greenwood, and were married in the living room, with Roy's cousin and the preacher's wife as witnesses. Then they drove straight to a motel to consummate the marriage.

"We left the motel late that afternoon," recalled Carolyn. On the road they passed her sister and brother-in-law going in the opposite direction. Both cars pulled over to the side of the road, and Carolyn told her sister the news. "My sister hugged us, wished us all the best, and hurried off to tell Mama." The happy couple jumped back in the car and drove on to Itta Bena, where Roy's family had gathered to celebrate the union; the decision to elope had been no secret among the Bryants and Milams. They enjoyed a huge supper and a festive evening, but Roy had to catch a bus back to Fort Bragg that night. "Here we were, married only a few hours, and he was going to leave me," she wrote later. "I was devastated."[11]

Carolyn found it bizarre that her mother-in-law, Eula Lee, drank whiskey for breakfast and carried a pistol in her purse at all times. Short, plump, and bossy, Eula Lee "could embarrass a sailor, cursing. . . . And could put away the booze. That's the first thing she did in the morning was fix herself a hot toddy. Bourbon." Her brother-in-law Melvin Campbell likewise poured down the whiskey. "All the time, from the very time he woke up in the morning until he passed out." Melvin in particular "could flare up in a minute. He had a real hair-trigger temper." Roy's brothers drank heavily, too, and were also quick to fly into a rage. "Well, it was like that with all of 'em. Roy was like that. That's the way every one of them was like."

Along with drinking hard, carrying a gun, and having a bad temper, overt expressions of white supremacy were simply part of the Milam-Bryant family's way of carrying themselves in the world. "They were racist, the whole family," confided Carolyn, implicitly exempting herself. "For one thing, it was the 'N-word' all the time. 'I've got this N working over here doing this, I'm gonna have to go get my money from that N over there because he's not paying me.'"

Her education in the ways of the Milam-Bryant clan began early in her marriage, at a Christmas gathering at her mother-in-law's store in Sharkey. "Most of us were back in the kitchen," she recalled, but Eula Lee, Roy, Mary Louise, and Melvin were standing at the front of the store. An old black man came into the store and said something to one of the women that displeased Melvin. "It was just because he either didn't say 'Yes, ma'am' or 'No, ma'am' or something." Melvin whipped out his pistol and swung it hard against the side of the man's head. Though he had only meant to use the .45 as a club, the gun went off and blew a hole through the wall of the store.

The roar of the gunshot "scared us all to death," recalled Carolyn, and the blow cracked the man's scalp. "Melvin hit him on the head and of course the blood splattered—you know how a head wound just bleeds so profusely." Mary Louise "had on a brand new white blouse, and she was standing close, and it just bloodied the front of her blouse. And as soon as we ran out there to see what was going on, here she is all bloody, and Melvin is holding the pistol, and we thought, 'Oh, Melvin shot Mary Louise!' And she was just laughing about it. They all were."

Carolyn's new husband didn't find the incident funny, she recounted. Instead, Roy expressed concern for the beaten and bleeding black man. What, Carolyn recalls he asked his mother, should he put on the man's scalp wound? Eula Lee advised turpentine. But first, Roy washed the gash out with soap and water. Only then, said Carolyn, did he pour turpentine over it, after which he borrowed his mother's truck and drove the poor black man home.

Here, drawn from Carolyn's memory, was the merciful gesture that never occurred to him a decade later when a fourteen-year-old black boy's life was at stake; here was the thing she wished he had done then. Apart from the details, the incidents bear an uncanny resemblance: an African American transgresses at the cash register, insulting a Milam-Bryant woman; ugly, quick, brutal violence rises up, policing that color line. But in this memory Roy responds with paternal concern; Roy becomes the responsible one. His brother-in-law Melvin "was very family-oriented, with his wife and children, but, well, he was mean as a snake, too."

History had stacked the social world of Jim Crow Mississippi like pancakes, with African Americans distinctly on the bottom. Pro-slavery ideologues of the late 1850s would have called black Mississippians the "mudsill," a foundational class to perform the necessary labors of life so that the higher classes could pursue the loftier aims essential to civilization.[12] One problem with this social structure was that middle- and lower-class whites tugged and scraped to find a satisfactory place for themselves. Their one undeniable accomplishment, which afforded a social status that could not be denied them, was to be born white. White sharecroppers, the lowest of whom even African Americans quietly dismissed as "poor white trash," occupied the rungs just above blacks. Laborers and small-time merchants like the Milams and Bryants, who made their living from selling cigarettes and snuff, illegal whiskey, and various snacks and staples, were only marginally higher; their betters derided them as "peckerwoods."

Though she liked to have a good time, Eula Lee was blunt. "If you [didn't] want to know the truth, [you didn't] ask her," her daughter-in-law chuckled. "The only thing I ever heard her say, actually, about [the Till murder] was that it was a shame I'd left the pistol in the car, I could have saved them all that trouble." Though tough as rough-hewn timber, Eula Lee pulled her children close to her and taught them to work hard and stick together. She organized an unflagging stream of family gatherings, often at her little grocery store in Swan Lake or the one after that in Sharkey, to eat big meals and drink a lot of whiskey. Somebody would bring fried chicken or pork chops. "It seemed like almost every weekend we were going to somebody's house or somebody was coming to our house," Carolyn recalled. "And we'd have just lots of regular old food, beans and peas and corn and potatoes, and all of it." There were few if any secrets. "Everything you did or got, whatever," Carolyn told me, "was everybody's business. And they were all in on it."[13]

The Milam-Bryant brothers were especially close, working together and regularly playing cards and drinking together. "You could never tell they were [only] half-brothers," said their mother, "unless they told you." Each of them carried a pistol. Seven of the eight of them had served in the military.[14] And most of them eventually ran small grocery stores throughout Leflore, Tallahatchie, and Sunflower Counties, in Swan Lake, Glendora, Minter City, Itta Bena, Ruleville, and Money. According to local law enforcement records, the stores sold whiskey in violation of the state's Prohibition laws.[15]

Eula Lee's sons practiced a raw style of masculine camaraderie that revolved around guns, hunting, fishing, poker, and drinking. Boys would be boys and women would stay out of it. Carolyn said, "They did such crazy things all the time anyway, you didn't really question what they were doing. 'Well, you wanna go see if we can find a deer?'—you know, at two o'clock in the morning—or 'Let's go get so-and-so and play some cards.' " They shocked young Carolyn with their loud arguments over practically anything. "They did have some of the worst arguments, cuss-fights, you ever heard, with their poker games. You would think they were gonna get up and fight, yeah." Eula Lee tried to reassure her daughter-in-law. "I would say 'Oh, my goodness' when I first got in the family, you know, and I thought, 'Ooh, what's going on, they're getting ready to fight in there.' And Mrs. Bryant would say, 'Oh, don't pay any attention to them, they're not gon' fight.' " Carolyn believed her, up to a point. "You just never knew what was coming—those kind of people. But they were *hard*."

The Milams and Bryants believed "that they could do anything they wanted to do and get away with it because they had a lot of clout"; that's how Carolyn put it years later. This confidence owed something to their association with the newly elected sheriff of Tallahatchie County, Henry Clarence Strider, known as "H.C." The sheriff was a 270-pound former football player who owned 1,500 acres of prime Delta cotton land and held sway over dozens of black sharecropping families. Carolyn described Strider as "sort of like the Godfather in Mississippi at that time. Whatever he said was what you did." Presumably because of their political support, the Milams and Bryants enjoyed his protection from the law and from everyone else. "One reason they were so much the way they were was that they thought they were in tight with him." Though Strider would prove to be a lordly ally, his vassals remained penniless peckerwoods.

As soon as Roy and Carolyn could scrape together the bus fare, she dropped out of high school and moved to North Carolina to be with him. She'd never left Mississippi before. Five months later she was pregnant and took the long bus ride back to her mother's house; the kindly driver pulled over time and time again to let his queasy young passenger throw up. In a few months she headed back to North Carolina with a three-month-old baby boy and got pregnant again almost immediately.[16]

When Roy was discharged from the army there was no question as to where the couple wanted to raise their growing family. They rented a little house in Glendora, where J.W. and his wife, Juanita, lived. It had front and rear screened porches like Aunt Mabel's house in Cruger. It helped to have someplace cooler to sit that first summer, when she was so hot and so pregnant. "I was pregnant with Lamar, and Juanita was pregnant with Harley [and] she would come over almost every day to help me with Roy Jr." Carolyn enjoyed and appreciated her sister-in-law, a quiet, soft-spoken woman whom photographers at the Till trial found quite fetching; "downright sexy" was how Carolyn remembered her. "When I had to go to the doctor, she would take me, or she'd bring her car and I'd drop her off and I'd go. I could depend on her." When Carolyn went into labor, Roy borrowed J.W.'s car to go to the hospital, where she gave birth to their second son.

Roy aspired to have his own trucking business, like J.W., so the young parents invested his army separation pay and her small savings to buy a dump truck. Roy began to haul gravel and make a little money. He soon hired and began to train another driver, but the trainee crashed the truck; both men escaped the flames, but the truck was a total loss. Worse, Roy learned that the insurance company he used had gone bankrupt just prior to the accident. Without funds to replace the truck, his dream was over, and he went to work for J.W.

One day in late 1953 Roy walked into their house and told Carolyn that they would be moving. With the help of J.W. he had bought a store in Money. Though he had rented the building, he had bought everything inside. Whether or not this was a good idea she did not know. "Roy never let me in on the planning of his business ventures with his family," she wrote. "I just accepted the fact that he was the breadwinner in our family, and [that] I needed to do as he asked me." Because of the store they would have most daily necessities close at hand, but otherwise they were still poor, unable to afford a car or a television set.[17]

Bryant's Grocery and Meat Market was on the first floor in front in a two-story brick building. The store became the local trading post for staples, tobacco, beer, snacks, and cold drinks, a place where people came to pass the time of day.[18] The post office, the filling station, and the store made up the business section of Money. The Bryants lived on the first floor behind the store.[19] The majority of their customers were black. Out front, under the awning, Roy and Carolyn had "made things fairly comfortable for the Negroes who patronize[d] their store," one journalist wrote, while whites chatted indoors. Several wooden benches and two or three checkerboards with bottle-cap checkers saw consistent use on the porch.[20]

Roy and Carolyn stocked the shelves with snuff, cigarettes, and cigars and the drink box with Cokes and grape sodas and R.C. Colas. In the glass-covered counter they put candy bars: Mounds, Baby Ruths, Kits, Paydays, Almond Joys, Butternuts, Hershey's bars, Butterfingers, and Milky Ways. There was penny candy, too—fireballs and Mary Janes, Milk Duds and peppermint sticks, Dum-Dums and Slow Pokes—and Juicy Fruit, Doublemint, Dentyne, and Double-Bubble gum. The cookie selections included Stage Planks, Jack's, Moon Pies, and all kinds of "Nabs," as they called Nabisco cookies and crackers and anything like them. On the counter sat a tall glass jar of pickled pigs' feet, another of big dill pickles, and a third of pickled eggs, with squares of wax paper to pick them up. They sold sardines, pork 'n' beans, milk, Wonder Bread, eggs, flour, lard, fatback, butter, bananas, cinnamon buns, Spic 'n' Span cleanser, and small household items of all kinds. The meat counter was in the back.[21]

If any of her children inherited Eula Lee's fiery temperament, it was the outsized middle child among the first set of five Milam children. He was the ruling body and spirit among all of them. "J. W. Milam was so domineering and so had to be in control all the time," according to Carolyn. Six feet two inches and more than 230 pounds, an extrovert and a decorated war hero, he had black hair around the edges of his massive bald head. "Big," they called him, because he took up a lot of space, literally and figuratively.[22]

Born in 1919 in Tallahatchie County, John William Milam attended school through the tenth grade. He served in the U.S. Army, 2nd Armored Division, from 1941 until the spring of 1946.[23] During the war he fought in Germany, often in house-to-house, hand-to-hand combat, and won a battlefield commission to lieutenant, a Silver Star, a Purple Heart, and several other medals.[24] "I do remember seeing J.W. with his shirt off one time, and he had some deep holes in his back, and when I asked Juanita about it, she said that's where the shrapnel hit him when he was injured in service," Carolyn recalled. His mother bragged, "He started as a private and got his commission the hard way." The journalist William Bradford Huie described Milam as "an expert platoon leader, expert street fighter, expert in night patrol, expert with a [Thompson machine gun], expert with every device for close-range killing."[25] His favorite weapon, however, was the army .45-caliber pistol. "I can tell you how good he was with that old pistol," his son boasted to FBI agents years later. "I seen him shoot bumble bees out of the air with it."[26]

J. J. Breland, a local attorney, regarded J.W. as a kind of brutal necessity for the social order of white supremacy. "He comes from a big, mean, overbearing family," Breland said bluntly. "Got a chip on his shoulder. That's how he got that battlefield promotion in Europe; he likes to kill folks. But, hell, we've got to have our Milams to fight our wars and keep the niggahs in line." One of four lawyers who would defend Milam and Bryant, Breland told Huie to let the country know that integration was out of the question in Mississippi. "The whites own all the property in Tallahatchie County. We don't need the niggers no more."[27] Of course this was hardly the case for the Bryants and Milams, who relied almost entirely on African Americans for their livelihood. Nor would they likely have agreed with Breland, despite their relative poverty, about their social position among white Mississippians.

Early in her marriage Carolyn clashed with J.W. on at least one occasion. "One thing that happened upset me pretty badly, and J.W. and I had a few words about it." The memory was sharp even sixty years later. "I said, 'I've got to figure out where to vote.' He says, 'Oh, you don't have to worry about that—I've already voted for you.' My first time to vote, and he voted for me." Her resentment still sounded fresh. "That's what he told me: 'You can't go back and vote, I've already voted for you.' I told him, 'Don't you ever do that to me again.' I'd been waiting all that time to vote and didn't get to."

If Roy objected to his half-brother voting for his wife, Carolyn did not remember it. J.W., fourteen years his senior, seemed to have a hold on Roy. When other people asked why they had different last names or referred to them as "half-brothers," Roy would say, "We don't consider us half-brothers. We're brothers." When Big Boy Bryant left the family, Big Milam took over the paternal role, especially for Roy.[28] The younger man certainly looked up to and took his cues from his hefty older sibling.

J.W. lived in the small community of Glendora, where he owned a store, a gas station, and a small house. He also owned a trucking business and an agricultural service firm; he had bought several large trucks and could repair and operate heavy farm machinery. When he had a long haul to make from Texas or Louisiana, he would often let Roy handle it.[29] He hired local African Americans to grade and repair driveways and gravel roads and take care of his trucks. They would also "wash his pickup and Juanita's car and stuff like that," his sister-in-law remembered. "Clean up, sweep out the store, take out the trash." Huie wrote, "Those who know him say he can handle Negroes better than anybody in the county." Milam himself agreed with the assessment, only adding the lie; "I ain't never hurt a nigger in my life."[30] Decades later Carolyn said of her brother-in-law, encompassing more than she meant and perhaps even knew, "He seemed to have a good relationship with them, and I think they were probably afraid of him too."

6

THE INCIDENT

On Wednesday evening, August 24, 1955, Reverend Moses Wright held services at the East Money Church of God in Christ. His three sons, Wheeler Parker, and Emmett Till had picked cotton throughout the morning and part of the afternoon. Summer was coming to an end, and soon enough they'd be parted for at least a school year. So Wright did not force the boys to attend church; instead he loaned them his 1941 Ford, instructing them to go no farther than the nearby country store because Maurice, who was driving, had no license. Six boys and one girl made the drive. The oldest was Thelton "Pete" Parker, nineteen, who lived nearby, as did Ruthie Mae Crawford, who was eighteen, and her uncle Roosevelt Crawford, fifteen. The two youngest were Simeon, twelve, and Emmett, fourteen. Wheeler and Maurice were sixteen. After Maurice dropped his parents off at the church, the group disregarded his father's imperative and drove to Bryant's Grocery and Meat Market in Money, only three miles away.[1]

When they arrived they gathered in front of the store; they'd come for candy and drinks but were in no hurry. "I noticed a group of young people congregating on the porch," Carolyn Bryant recalled. "It was not unusual for people to be on the porch as there were many checker games played by the locals outside our window." "Locals" was a courtesy used decades after these events. At the time, in polite company, at least, they were called "Negroes," and the store survived because they gathered on that porch "almost daily."[2]

As they talked that day it is possible that Emmett claimed he attended school with white girls and even claimed he had dated some of them. One cousin recalled that Emmett had a picture of a white girl in his wallet and showed it to the others, bragging that it was his girlfriend. Mamie Bradley later wrote that Emmett's wallet had come with a picture of the movie star Hedy Lamarr in it, although she may have been trying to defend her son's reputation against any imputation that he had a sexual interest in white girls, which would have undermined public perception of his innocence.[3] None of the witnesses interviewed soon after the incident mentioned either a photograph or any braggadocio from Emmett, so these may be embroidered tales. There is some agreement among the witnesses that an unidentified boy, not one of those who came with the group, suggested that Emmett go inside the store to at least look at Carolyn Bryant. Wheeler, for instance, told a reporter, "One of the other boys told Emmett there was a pretty lady in the store and that he should go in and see her."[4] Wheeler went in first, and Emmett was not far behind, intending to buy bubblegum. Wheeler made his purchase quickly and left, leaving Emmett in the store. "For less than a minute he was alone in the store with Carolyn Bryant, the white woman working at the cash register," Simeon wrote. "What he said, if anything, I don't know."[5]

Nor do we. On September 2, when Carolyn told her attorney in private what had happened at the store, she claimed only, "I waited on him and when I went to take his money, he grabbed my hand and said how about a date, and I walked away from him, and he said, 'What's the matter, baby, can't you take it?' He went out the door and said, 'Goodbye,' and I went to the car and got [the] pistol and when I came back he whistled at me—this while I was going after [the] pistol—didn't do anything further after he saw [the] pistol." It is crucial to note that this account does not include the physical assault she testified to in court twenty days later.[6] The *Greenwood Morning Star*, having interviewed Sheriff George Smith of Leflore County, reported on September 1, "The Bryants were said to have become offended when young Till waved to the woman and said 'goodbye' when he left the small store Saturday night."[7] Two days later Smith added the statement "Till made an ugly remark to Mrs. Bryant."[8] Another report indicated that Emmett was "insolent" to Carolyn, having failed to say "Yes, ma'am" when addressing her.[9] When J.W. and Roy invaded the Wright home with guns to kidnap Emmett, they called him "the one that done the *talking* up at Money" and "the one that done *the smart talk* up at Money," so clearly they were focused on a verbal offense. If Emmett had put his hands on Carolyn, it is unimaginable that neither J.W. nor Roy would have mentioned it. "The talking," "the smart talk," "an ugly remark"—at that point no one mentioned or even suggested that anything physical, sexual, or threatening had happened.

Years later Ruthie Mae Crawford told the documentarian Keith Beauchamp in an interview that she watched Emmett through the plate glass window the entire time. She insisted that the only mistake he made was to place his candy money directly in Carolyn's hand rather than put it on the counter, as was common practice between whites and blacks.[10] This alone would have violated Mississippi's racial etiquette. How serious a violation was entirely a matter of mores, not a question of law. Of course, even to look Bryant directly in the eye would be to break the "cake of custom" that accompanied legal segregation.[11]

In an interview in 2005 Simeon supported Ruthie Mae's recollection, telling a reporter that Emmett had paid for some gum and placed the money in Bryant's hand, violating a Mississippi taboo about which he may have known nothing.[12] Whatever Simeon saw propelled him to act: he quickly went into the store, perhaps to extricate Emmett from potential trouble. "After a few minutes," wrote Simeon, "he paid for his items and we left the store together."[13]

Wheeler said Emmett was in the store alone with Carolyn for "less than a minute"; Simeon measured it at "a few minutes." What could possibly have transpired in such a short time? Between "within a minute" and "after a few minutes" is the time frame we must fill. Between Parker and Till entering the store, between Parker leaving his cousin alone in the store and Wright and Till leaving the store, something happened that Carolyn would decide deserved, what? Something. Whatever it was made Carolyn angry enough to fetch a pistol. On the way out of the store, Maurice Wright reported, Till turned and told Carolyn Bryant, "Goodbye."

Before the seven black youths dispersed, Carolyn stomped out of the store, heading straight toward her sister-in-law's car, which was parked near the door. If she had been afraid, if what she had experienced had been tantamount to sexual assault, as she testified in court, walking through the group outside would have been an odd choice; she could just as easily have turned the deadbolt on the front door and stayed inside. It seems whatever happened in that store made her more mad than fearful.

"She's going to get a pistol," Wheeler reported one of the boys saying.[14] Carolyn reached under the driver's seat for her husband's .45 automatic. All agreed that at this point Emmett let out a "wolf whistle." "To this day I don't know what possessed Emmett to do that," Simeon said. "We didn't put him up to it. Many of the books and stories said that we dared him to do it. But that's not the truth. He did it on his own, and we had no idea why."[15]

"He did whistle," Carolyn told me, "but it was after he'd been in [the store] and I went out to get the pistol."

Simeon wrote later, "We all looked at each other, realizing that Bobo had violated a longstanding unwritten law, a social taboo about conduct between blacks and whites in the South." The local youths were dumbfounded. "Suddenly, we felt we were in danger and we stared at each other, all with the same expression of fear and panic. Like a group of boys who had thrown a rock through somebody's window, we ran to the car."[16]

The old Ford raced down the blacktop, but any relief the young people felt evaporated when one of them looked in the rearview mirror and saw headlights coming up fast behind them. Convinced that they were being followed, they decided to pull over and flee. Scattering in all directions through the dark cotton fields, they stopped only when the car flew past without even slowing down. Laughing a little, they decided their fear had been paranoia and piled back into the Ford to head home. They edged closer to the judgment Carolyn Bryant would reach decades later: that nothing their cousin had done at the store amounted to much.

Even so, as they drew near the Wrights' place, Emmett begged his cousins not to tell his uncle and aunt about the incident. They all agreed, fearing that if the grown-ups heard about it Emmett might end up on the train to Chicago earlier than anyone had planned.

Someone did tell Elizabeth Wright, however, and her husband soon heard it, too, though who told them is not clear. Word was spreading. The next evening, Simeon wrote, "a girl who lived nearby told us she had heard about what happened in Money and that trouble was brewing. 'I know the Bryants, and they are not going to forget what happened,' she warned us."[17]

7

ON THE
THIRD DAY

Just after Emmett was kidnapped, Elizabeth Wright ran to the home of white neighbors to ask for help. It was not yet three in the morning. The darkness was complete, but urgency guided her steps across the fields to the distant house where the farmer and his wife slept. Perhaps she was too timid to explain herself well. Perhaps a black woman wailing at the door about a fourteen-year-old boy abducted by white men with guns frightened the couple. However Elizabeth asked for help, the wife wanted to give it but the husband would not consent, and Elizabeth returned home in tears, vowing to leave Mississippi forever. She insisted that Moses take her to her brother's house in Sumner, so, leaving the five boys in their beds, they drove about half an hour to the home of Crosby Smith.

Elizabeth barely paused in her flight north. She stayed in Sumner a few days until she rode with Emmett's body back to the train platform in Chicago, faced with explaining all this to his mother. She would never return to her house in Mississippi.[1] Moses received a letter from her, from Chicago, dated August 30, two days after the kidnapping. "Come up here," she wrote, "and tell Simeon to get my corset and one or two slips and a dress or two and bring them to me. If Eula sent my dress, bring it, also my stockings. I mean you come."[2]

On the way back from the train station Moses picked up his brother-in-law Crosby Smith before driving back to Money. He intended to speak directly to the men who had threatened his life and kidnapped his nephew and ask them to return the boy—if he was still living. At about four or five in the morning, on a deserted street, the two men knocked at the back door of Bryant's Grocery and Meat Market. Not knowing if they would be met with success or a shotgun blast, they did so with less noise than Roy and J.W. had made on the Wrights' porch hours earlier.

"I heard a knock on the door," Carolyn remembered. It was not loud. She was alone with the children, and, fearful about who would call on her at such an hour, she kept silent. "I heard another voice, a different voice. The voice sounded like a black man, saying to another person, 'It looks like there's nobody here.' I heard a car slam, then the sound of a car driving away." The men drove to the Wrights' house in East Money to wait for daylight.[3]

At about nine on the morning of the kidnapping, Curtis Jones, one of the three cousins from Chicago who were staying with the Wrights, went to a neighbor's house and called his mother, Willie Mae, to tell her the terrible news. She called Emmett's mother: "I don't know how to tell you. Bo." Mamie's mind began to race in horror. "Some men came and got him last night." Mamie drove to her mother's apartment at breakneck speed to share both news and anguish. Her mother alerted their church so that the congregation could pray for Emmett. Then Mamie did a curious thing, foreshadowing the boldness with which she would handle this tragedy. She started telephoning Chicago newspapers, which soon sent reporters to Alma's apartment.[4]

It is important to recall what was and wasn't possible in 1955. Mamie surely knew there was no existing local authority in Mississippi she could rely on, nor was there any federal authority she could reasonably hope to enlist. One exception was the press, particularly the black newspapers of Chicago. At some point, too, she called Rayfield Mooty, a distant relative and a savvy labor union official who knew all the African American political players in Chicago; he would become a key advisor. When he got to her mother's apartment, Mamie told him, "Mr. Mooty, we just want you to take the case over. Daddy got a lot of confidence in you and he says you can do it."[5] While the fullness of her plan probably hadn't taken shape in her mind, Mamie was already determined that they would not proceed quietly.

Watching the morning rise around his fields and porch, Moses Wright still hoped that Emmett would come back alive, that the white men would think better of killing him and bring him home. After a couple of hours, though, he and Crosby Smith drove to the office of Sheriff George Smith of Leflore County. Sheriff Smith knew Roy and J.W. well and immediately assumed that they had killed the boy and thrown his body in the river.[6] The three men left to search for clues: Crosby Smith rode with one of the deputies, while Moses joined Sheriff Smith. "We looked under many a bridge that day," Crosby Smith recalled. "Because that's the first thing Moses thought. It was custom, what was being done around here in those days. We went by custom when something like that happened, and that's usually what they done to 'em."[7]

They found nothing. Just before two o'clock they called off the search, and the sheriff left to question Roy. When Moses got home, well-meaning visitors were waiting at the house. "By Sunday noon, every Negro for miles around knew all about it," he said. "The people kept coming and we prayed and prayed." Rumors began to spread that J. W. Milam and Roy Bryant were the ones who had killed Reverend Wright's nephew from Chicago.[8]

While the Wright household prayed that Emmett was still alive, Roy Bryant slept. As Carolyn describes the scene in her memoir, when Roy came home in the early morning light, she asked him where he had been and what he and J.W. had done with the boy. This scenario supports later testimony that the brothers brought Emmett to the store for her to identify. However, it does not close the speculation about whether it was indeed Carolyn in the darkened vehicle hidden on the Wrights' property at two in the morning who identified Emmett as the boy who had assaulted her.

Roy assured her that he'd been playing poker all night at J.W.'s store in Glendora with his brothers and Melvin Campbell and a few others. According to Carolyn, he insisted, "We just whipped the boy and then dropped him off on the side of the road. That was it. We didn't do anything else to him." She maintained even years later that she had believed him at the time.[9]

Roy was still asleep when Sheriff Smith parked his squad car in front of Bryant's Grocery at two in the afternoon. He woke only when he heard the pounding on the back door; then he quickly dressed and stepped outside, where the sheriff and his deputy, John Ed Cothran, were waiting. Smith and Roy got into the police car to talk. "I asked him about going down there [to the Wrights' house] and getting that little nigger," Smith testified. Roy acknowledged that he had done that. "I asked him why did he go down there and get that little nigger boy." Roy said he had heard that the boy made some "ugly remarks" to his wife. He "said that he went down and got [Emmett Till] to let his wife see him to identify him, and then he said she said it wasn't the right one, and then he said that he turned him loose." Smith asked where they'd released Till. "He said right in front of the store. He said he went to some of his people—I don't remember who he said just now—and he said he played cards there the rest of the night." The sheriff arrested Roy for kidnapping and booked him into the Leflore County Jail.[10]

The rest of the country did not even know Emmett was missing—not yet. Mamie was only just alerting newsmen in Chicago. Moses spirited his grandson Wheeler to Greenwood to catch the train home to Chicago, which left him with three sons and one grandson. He still had twenty-five acres of cotton to pick and sell, money they would need under any circumstances, so the boys would help him.

Sheriff Smith continued to search for the corpse and seemed determined to build a case that would yield convictions. On Monday, August 29, the day after the sheriff arrested Roy, the Bryant-Milam clan gathered at Eula Lee's store in Sharkey to discuss the situation. It was terrible that Roy had been arrested, of course, but now they wanted to make sure things didn't get out of hand and sweep up the entire family. Roy was not the strongest stick in the bundle, and there was some concern that he might break under pressure and implicate everyone. It was decided that J.W. would allow himself to be arrested in order to keep Roy from "running his mouth" and changing the story they had agreed to tell. J.W. drove straight to the sheriff's office, where he was immediately arrested for kidnapping. He acknowledged that he had abducted the boy but claimed to have released him that night, and he refused to implicate anyone else, even Roy. Other than that, J.W. didn't say a word.[11]

On Wednesday, August 31, the third day after the kidnapping, Emmett's body surfaced through the dark water. "I seen two knees and feet," said the boy who discovered it. Robert Hodges, seventeen, was walking along the riverbank just after dawn. The ruddy-faced sharecropper's son kept several "trot lines" stretched across the mud-brown Tallahatchie, and he was hoping to find them thrashing with one of the big catfish that lurked in the cool, dark depths. But this morning, as the dawn seared the mist off the river, the boy saw toes protruding from the water: "[The body] was hung up there on a snag in the bottom of the river."[12]

Decades later Carolyn Bryant still recalled the news about the body as a traumatic shock. After Roy's arrest she and her two sons were taken in by the Milam and Bryant relatives. "I suppose they contacted Leslie or Melvin and they made arrangements," she told me. The family kept her isolated from the unfolding aftermath of the lynching and would not even let her speak to anyone on the telephone. On the third day Carolyn was at Buddy Milam's store when Raymond, her husband's twin brother, walked through the door. "They found a body this morning," he announced. "They think it's [Emmett Till]."

"No, it can't be," she said. "Roy told me he didn't do anything to him, that they turned him loose."

"And [Raymond] said, 'Well, Roy didn't do it. Melvin Campbell did it.' And that's when I told him, 'Well, why is Roy and J.W. in jail and Melvin's out? Why would they arrest Roy, then?' And he said I was not to tell anybody it was Melvin. And I guess Raymond told [the rest of the family] what I had said because they got us that night and took us to a sister-in-law's house in Lambert, Mississippi. And we were really isolated then."

The farmhouse in Lambert was about a forty-five-minute drive from the Bryant store in Money. Carolyn and her sons stayed there only a couple of days, and then the Milam-Bryant clan sent yet another relative to move them to another safe location. "I don't remember who got us but it was always one of [the Milams]. Me and my two boys, we'd stay here a couple of nights, and then they'd take me to another relative and we'd stay a couple of nights." The Milams refused to let her telephone even her mother or siblings, who became alarmed when they did not hear from her. "My brother came looking for me, and one of the Milam brothers told him that I didn't want them to know where I was, that I didn't want to talk to them," Carolyn told me, still sounding offended. "They were keeping me from everybody. They were afraid that I might say something they didn't want me to say or I might reveal something they didn't want revealed."[13]

Sheriff Smith told reporters for the *Greenwood Morning Star* on August 30 that he wanted to bring her in for questioning. She ought to know something useful; after all, it was Emmett Till's "alleged insulting remarks" to her that provoked the kidnapping in the first place.[14] The following day, the same day Emmett's body was found, the *New York Post* reported that Leflore County authorities issued a warrant for the arrest of Carolyn Bryant on kidnapping charges.[15] The *Chicago Daily Tribune* reported that the Sheriff's Department had been unable to locate her to serve the warrant.[16] Two days later, strange as it seems, the *Birmingham* (Alabama) *News* announced, "Authorities apparently have abandoned the search for Mrs. Roy Bryant." Leflore County district attorney Stanny Sanders had told the

THE BLOOD OF EMMETT TILL

Birmingham reporters that there were "no plans at present for picking up Mrs. Bryant."[17] Fifty years later the FBI informed Carolyn that a warrant had been issued for her arrest; that, she wrote in her memoir, was the first she'd heard of it.[18]

A spokesman for the Leflore County Sheriff's Department told the *Memphis Commercial-Appeal* on September 2, "We know where she is and feel sure we can pick her up if needed."[19] But by the next day Sheriff Smith appeared set on leaving Carolyn out of it entirely. He claimed a chivalrous impulse that apparently overrode the belief that she had participated in the kidnapping. "Officers have not questioned Bryant's pretty brunette wife, in her early 20s, who was believed to have stayed in the car with an unidentified man when Bryant and Milam whisked Till from the home of his uncle in Money, Miss. community," stated the *Greenwood Morning Star*. "We aren't going to bother the woman," Smith told reporters on September 3. "She has got two small boys to take care of."[20]

Despite the unsettling sight of human toes protruding from the water, Robert Hodges finished checking his fishing lines, then hurried home to find his father, who reported what Robert had seen to B. L. Mims, their landlord. Mims called his brother, Charles Fred Mims, then Deputy Sheriff Garland Melton of the Tallahatchie County Sheriff's Office. The deputy called Sheriff H. C. Strider, who telephoned his teenage son. "My dad called and asked me did I have a boat in the river," Clarence Strider Jr. recounted, "and I told him I did. Then he said we'll be down there in a little while and he sent the deputies to go with me."[21]

Deputy Melton and Deputy Ed Weber picked up young Strider and made their way to the landing at Pecan Point, twelve miles north of Money, where they met Robert Hodges and his father. They took both boys' boats into the muddy river and quickly saw what Robert had seen. "Well," B. L. Mims said, "we saw a person—from his knee on down and including his feet—we saw that sticking above the water. And we could tell by looking at it that it was a colored person." With the engine humming low, they navigated over to the body. Something was holding it head downward in the water. The men tied a rope around the legs and used the motor to pull the body upriver a few feet, until the weight holding the head down came unsnagged from whatever pinned it to the bottom of the river. Towing it over to the landing, they dragged the corpse up onto the bank. There they could see that an iron fan, the kind used to ventilate cotton gins, was lashed to the corpse's neck with several feet of barbed wire. Packed with mud from the bottom of the river, the fan weighed about 150 pounds. Whoever had disposed of the body had intended that it never rise from the river bottom.[22]

As the men examined the body, Deputy Sheriff John Ed Cothran of Leflore County arrived, as did Sheriff Strider of Tallahatchie County. Strider noted what looked like a bullet wound above the right ear and that the other side of the boy's face was "cut up, pretty badly like an axe was used." That left side "had been beat up or cut up—plumb into the skull." The sheriff estimated that the bloated body had "been in the water about two days."[23] He dispatched Cothran and another deputy to Money to fetch Moses Wright to identify the body. Cothran also contacted Chester Miller, a black undertaker at Century Burial Association in Greenwood, who, along with one of his assistants, met them down at the river.

Miller testified that when he approached the body, it was lying face-down in a boat. Turning it over, he saw what looked like a heavy wheel attached to the corpse with a strand of barbed wire that "was well wrapped" around the neck. He unwound the wire to remove the fan. Miller turned to Moses Wright and asked, "Will you identify the body as the boy who was taken from the house?"[24]

"I was standing right up over him," Wright recalled. "They turned him over and then I saw all of it." He nodded at Miller, who saw no reason why Wright should have any trouble identifying the body, even in its state of considerable damage and decomposition. Deputy Cothran told reporters soon afterward that Wright "definitely identified the body as the boy." The law officers asked Miller to remove the silver ring from Till's finger. Miller gestured to his assistant, who was wearing rubber gloves. "Well, he had the gloves on," Miller explained, "and then I said to him, 'Take [the ring] off.' And then he took it off and handed it to me. I laid it on the floorboard of the ambulance," referring to the hearse he had driven to the river. Miller noted that the ring was engraved with the initials "L.T."[25]

As he prepared to transfer the body to a casket, Miller was stunned at the extent of the wounds. "The crown of his head was just crushed out and in, and a piece of his skull just fell out there in the boat, maybe three inches long [and] maybe two and a half inches wide, something like that." There was a hole perhaps half an inch square above the right ear, which Miller assumed was a bullet hole. He and his assistant placed the body in the casket and hefted the casket into a metal shipping case, which they then pushed into the hearse and drove to the Century Burial Association.

Given the state of the body, there was little magic that the undertaker's art could perform; a closed-casket funeral seemed certain. An officer from the Greenwood Police Department took pictures of the corpse. Miller assumed that the family would contact him about the funeral, but Sheriff Strider phoned Miller with an unusual demand: he should bring the body to Reverend Wright's Church of God in Christ in East Money for burial *that very day*. Apparently Strider wanted no one to see the condition of the corpse. Miller did as he was told: "I delivered it to the cemetery at Money."[26]

Strider announced his jurisdiction over Milam and Bryant's case even before murder charges had been filed against them. Although the kidnapping had occurred in Money, which is in Leflore County, and no one knew where the boy had been killed, Strider maintained that the body was put into the river "a good 10 miles" into Tallahatchie County and must have been dumped there since "it couldn't have floated up the river."[27] District Attorney Sanders sided with Strider: the youth was abducted in Leflore County but the body was recovered in Tallahatchie County, thus giving Tallahatchie jurisdiction for prosecuting the case.[28]

His authority over the case confirmed, Sheriff Strider determined that the body would be buried immediately. Charged with investigating a presumed murder, he saw no reason for an autopsy. They needed to get on with it, he thought. Nobody needed to see this body. And so someone

notified a few of Till's Mississippi family members that the body was to
be buried right away and that they might want to be present. That they
immediately complied probably illustrates the extent to which it was not
safe for African Americans to challenge white men in 1950s Mississippi,
which was doubly true if the white man in question was Sheriff H. C.
Strider. Moses Wright began preparing his eulogy, and others readied their
funeral clothes.

When Curtis Jones, Emmett's Chicago cousin, saw preparations for
burial proceeding, he called his mother in Chicago, who notified Mamie.
She was incensed. She had already decided to hold the funeral in Chicago.
She called her uncle Crosby Smith in Sumner, who promised "to get Emmett's body back to Chicago if he had to pack it in ice and drive it back
in his truck."[29] He drove to the cemetery with Chester Miller and one
of the sympathetic Leflore County deputies. "[The grave diggers] had got
the body out to the cemetery and dug the grave," Smith recalled. "I got
there and had the deputy sheriff with me. He told them that whatever I
said, went." Smith told the grave diggers, "'No, the body ain't going in the
ground.' That body went to Chicago."[30]

Here was another moment when the Till case could have become just
another private bereavement and another mother's appeal to her church
and the local press for justice. Had his body been buried hastily in Mississippi soil, most of the rest would almost certainly not have followed.
There would have been a trial and some outrage, but without the Chicago
funeral, would the rest of America, let alone the world, have paid attention?

Miller, who had already hauled the corpse from the river to the funeral home in Greenwood and from there to the cemetery in Money, got
Smith to help him trundle the coffin back into his hearse and headed
back to Greenwood. However, he told Wright, "I don't dare let that body
stay in my establishment overnight." He considered it very bad luck, to
put it mildly, to defy Sheriff Strider. "I wouldn't have any place in the
morning and perhaps wouldn't be alive by morning." Wright then telephoned a white undertaker, C. M. Nelson, in Tutwiler, forty miles west

of Greenwood, and asked him to pick up the body. Nelson owned two funeral homes: one for blacks and one for whites. A man of considerable wealth, he also served as mayor of Tutwiler.

Nelson agreed to prepare Till's body under one condition: Wright had to promise that the seal on the casket would never be broken and that no one would ever be allowed to view the body. Wright agreed, making a promise he'd be powerless to keep, but events were moving quickly in a new direction. The badly swollen body prevented intravenous embalming, so to ensure at least minimal preservation Nelson's embalmer immersed it in a vat of formaldehyde and made incisions to release the tissue gas.[31]

Thanks to Mamie, Chicago's newspapers, radio, and television were already starting to cover the lynching. A TV news bulletin even interrupted *I Love Lucy* to report the discovery of the body. Now word spread that Emmett Till's body was coming home to Chicago. Mamie now envisioned God's purpose for her life—and for her son's life: "I took the privacy of my own grief and turned it into a public issue, a political issue, one which set in motion the dynamic force that ultimately led to a generation of social and legal progress for this country."[32] Unlike any of the white newspapers, soon after Till's lynching the *Pittsburgh Courier* predicted that his mother's "agonized cry" might well become "the opening gun in a war on Dixie, which can reverberate around the world."[33] Activists across the country hoped and believed that this tragedy might be the wellspring of positive change. Mamie had ensured that to her mother's cry would now be added the mute accusation of Emmett's body.

A colleague wrote to the activist Anne Braden soon after the lynching, "This 14-year-old's crucifixion is going to strengthen and clarify the cause of de-segregation, human brotherhood, and freedom."[34] It would fall to Mamie Bradley to transform crucifixion into resurrection.

8

MAMA MADE THE
EARTH TREMBLE

On Friday, September 2, 1955, Emmett Till's mother focused much of black Chicago on her son's murder and the movement it could help unleash. "By the time we reached the train station at Twelfth Street early that Friday morning there was already a huge crowd," Mamie wrote years later. A thousand people packed the platform. "I had to be brought up in a wheelchair. I was too weak and I just couldn't stand up at the moment the train pulled in."[1] Reporters and photographers from virtually every Chicago newspaper recorded the scene. The *Chicago Defender* reported, "Limp with grief and seated in a wheelchair among a huge crowd of spectators, Mrs. Bradley cried out: 'Lord you gave your only son to remedy a condition, but who knows but what the death of my only son might bring an end to lynching.'"[2] The *Chicago Sun-Times* described a "hysterical scene" after the train bearing Till's body arrived: "Mrs. Mamie E. Bradley jumped from her wheelchair Friday when the

Illinois Central Railroad's Panama Limited pulled in. She sprinted across three sets of tracks to the baggage car in which the body lay in a pine box." Sobbing wildly, she fell to her knees. "My darling, my darling, I would have gone through a world of fire to get you." With weeping relatives forming a ring around her and the hearse backing into the scene, Mamie yelled again, "My darling, my darling, I know I was on your mind when you died." As stevedores lifted the pine box into the hearse, the *Sun-Times* reported, "she said softly: 'You didn't die for nothing.' "[3]

The uneasiness she had felt about Emmett going to Mississippi, the fear that gripped her when she heard that armed white men had snatched him away, the horror when the worst that could happen became undeniable fact, all began to flow out of her at once. "And I kept screaming, as the cameras kept flashing," she wrote, "in one long, explosive moment that would be captured for the morning editions."[4]

That sentence encapsulates the next several months of her life.

Uncle Crosby Smith, who had accompanied the corpse from Mississippi, stood alongside her on the platform, as did her sweetheart, Gene Mobley, and Rayfield Mooty, now more or less her political advisor. Smith reportedly took Mooty aside and urged him, "Don't let nobody see that box, don't even let them open the box. Be sure don't let Mamie see what's in there."[5] Mooty stayed close at hand, as did Bishop Louis Henry Ford and Bishop Isaiah Roberts, who pushed her wheelchair and prayed with her. The sight of the stevedores hefting that huge pine box and rolling it toward a waiting hearse made her stand up and then fall to her knees. The two ministers laid firm hands on her shoulders. "Lord, take my soul, show me what you want me to do," she cried, "and make me able to do it."[6]

Mamie was far from the first American mother to cry bitter tears over a child lynched in Mississippi. On October 12, 1942, according to an NAACP investigatory report, two fourteen-year-old black boys, Charlie Lang and Ernest Green, were seen by a passing motorist playing with a white girl near a bridge and charged with attempted rape. A mob seized them from the jail in Quitman, cut off the boys' penises, and pulled chunks of flesh from their bodies with pliers. One of the boys had a screwdriver shoved down his throat until it protruded from his neck. The mob then hanged the boys from the bridge, a traditional lynching site in Clarke County. A photograph of their bodies taken surreptitiously was released by national wire services, but only one white newspaper, *PM*, printed it, though a number of African American papers did. The *New York Times* reported the lynchings without the photograph in a one-column story on page 25.[7]

Only weeks before the Till lynching, terrorists had assassinated Reverend George Lee and Lamar Smith for their efforts to register black voters. The national press and the federal government ignored the murders, seeming to accept that this kind of behavior was a fact of life in the Magnolia State and a matter of little concern outside activist circles in the rest of America. "Emmett Till was, you know, that sort of a strange phenomenon," Clarksdale NAACP leader Aaron Henry told an interviewer in 1981. "White folks have been killing black boys all of my life, throwing them in rivers, burying them, and all that shit. Just why the Emmett Till murder captured the conscience of the nation, I don't know. It could have been that it was the beginning of television and people could see things. The fact that a black boy was killed by white men wasn't nothing unusual."[8]

Emmett's murder would never have become a watershed historical moment without Mamie finding the strength to make her private grief a public matter. Today was the day he should have bounded from the train with stories of the Delta to tell, eager to know if she'd gotten his bicycle fixed, ready to take to the sandlots and dream he was Don Newcombe pitching Brooklyn into the World Series. He should have bounced into the house eager to see his dog, Mike, rescued from the pound. He was supposed to start school in a few days, and finish painting the garage door.[9] But now his mother's errands included finding a way to bury her own heart, finding a way to go forward without a reason to do so, and giving her only son back to God. She would do all that, and she would leverage the only influence America's racial caste system granted her: public grief and moral outrage sufficient to shame and anger some fraction of the nation.

•   •   •

Rayfield Mooty claimed that he urged Mamie to open the casket and keep it open.[10] But the choice to display the ruined body of her son was Mamie's alone to make, and there is no reason to doubt that she made it deliberately. Even before they left the train station, the funeral director A. A. Rayner, who had arranged for the hearse that would carry Emmett's body to the South Side funeral home, had clashed with Mamie over her determination to open that box. Rayner had agreed not to open it before the coffin even left Mississippi; she insisted that he do just that.

In the time-honored way of undertakers, Rayner remained serene. He had served the South Side for years and knew that many funerals generated brief squalls about what was proper and best. Grief set people on edge; the funeral director's job was to absorb the tension calmly and yet make sure that things were done right—in this instance, that the box stay shut. "He was very patient with me," Mamie recalled. "He set it all out for me. It was being sent locked up with the seal of the State of Mississippi."

"Mrs. Bradley," he explained, "I had to sign papers, the undertaker had to sign papers, your relatives had to sign papers." A number of promises had been necessary to get the casket out of Mississippi, the main one being that it would never be opened. No one needed to see what was in that box anyway, he told her.

"I told him that if I had to take a hammer and open that box myself, it was going to be opened," she wrote. "You see, I didn't sign any papers, and I dare them to sue me. Let them come to Chicago and sue me." Finally Rayner relented.[11] Before leaving the train station either Mamie or one of her companions told Simeon Booker, a reporter for *Jet* magazine, they were going to A. A. Rayner & Sons, and Booker instructed a photographer to follow him there. "Mrs. Till didn't have anybody else in the press she knew," he recounted.[12]

As they neared the funeral home at 41st Street and Cottage Grove, Mamie and her entourage could already detect a ghastly odor. Inside the staff set off spray canisters of deodorizer to cover the smell, though the attempt was unsuccessful. Rayner showed them to a waiting room so his staff could have a little more time to prepare the body, but Mamie demanded to see her son's body as it was. Reluctantly Rayner led Mamie, her father, and Gene Mobley to the room where Emmett's corpse lay on a slab for embalming. "He really didn't think that I should look at Emmett like this," she recalled. "But I had kept insisting."[13]

Her two companions steadied her on either side. "At first glance," she wrote, "the body didn't even appear human." But she recoiled in horror from the realization that "this body had once been my son." She held herself closely in check, trying hard to "steel [herself] like a forensic doctor. I had a job to do." She began with the feet, noting the familiar ankles, legs, and the knees so much like her own. There were no noticeable scars on the body, but the huge tongue seemed choked from his mouth. His right eyeball rested on his cheek, hanging by the optic nerve, and the left eye was gone altogether. The bridge of his nose seemed to have been chopped with a meat cleaver, and the top of his head was split from ear to ear. A bullet hole just behind his temple showed daylight from each side. The vision was ghastly, but she had done her duty: it was Emmett's body.

"That's Bobo," Gene Mobley, the barber, said. "I know that haircut."

Mamie's last thoughts as she turned away were how her boy must have felt that night when he realized that he was going to die. She knew in her heart that he must have called out her name.[14]

At the funeral home, according to the *Chicago Sun-Times*, Mamie made up her mind to have an open-casket funeral, exclaiming, "Let the people see what they did to my boy." She dispatched Mooty to talk with her minister about using the church. Mooty recalled telling the man, "All we want is to have a place to lay that body where it can be seen, that's all. This may not stop [lynching] but this will be starting an end to all that lynching that's been going on down in Mississippi. They've been throwing everything in the river until it be full. Now you got a chance to be a great man if you let us use that church." Whatever convinced the preacher, the church doors opened for the viewing of the body that very night.[15]

When Mamie informed Rayner that she wanted the casket open for the viewing and the funeral ceremony, he did not try to dissuade her. Instead he asked if she wanted him to retouch Emmett's body and make him look a little more presentable. "'No,' I said. That was the way I wanted him presented. 'Let the world see what I have seen.'"

Rayner took the liberty of slightly preparing the body anyway. He stitched the mouth closed to cover the ghastly tongue, removed the dangling eye, and closed both sets of eyelids. On the left side of the head, where the beating had been most severe, he sewed the pieces of the skull back together as best he could. The body remained a horrifying sight. Mamie remained gracious about this slight subversion of her will: "I told Mr. Rayner he had done a beautiful job."[16] Inside the coffin Rayner laid the body in a glass-covered, airtight box that successfully contained the horrific odor.

That evening, at the first public viewing, many thousands stood in long lines around the block to pay their respects and see Emmett Till's mangled body; the *Chicago Defender* reported "more than 50,000," though estimates varied.[17]

Minnie White Watson, an archivist at Tougaloo College in Mississippi, believed that her friend Medgar Evers, the state NAACP field secretary, was instrumental in "talking Mrs. Till into having . . . [an] open casket funeral" for her son. Mamie later acknowledged that she was "grateful for his commitment and his compassion. He had really been moved by Emmett's murder. He was the one who had done the initial investigation to brief the NAACP head office."[18] Evers and his allies in Mississippi had displayed George Lee's disfigured face in an open casket to great effect. Though Mamie's memoir does not exclude the possibility of Evers's influence, it eloquently describes how she came to the decision:

> I knew that I could talk for the rest of my life about what happened to my baby, I could explain it in great detail, I could describe what I saw laid out there on that slab at A. A. Rayner's place, one piece, one inch, one body part at a time. I could do all of that and people would still not get the full impact. . . . They had to see what I had seen. The whole nation had to bear witness to this. I knew that if they walked by that casket, if people opened the pages of *Jet* magazine or the *Chicago*

Defender, if other people could see it with their own eyes, then together we would find a way to express what we had seen.[19]

Mamie and Ray Mooty continued to work the phones and keep the case before the Chicago media. Had they not done so it would be difficult to explain the *Chicago Tribune*'s report the following day, September 3: "More than 40,000 persons viewed the body in the afternoon and night." The line stretched around the block and beyond throughout the day and into the evening. So many people became overwhelmed by emotion at the sight of the body that the funeral home had to set up a special section of chairs outside where they could sit and recuperate. Ushers and women in white stood ready to catch those who fainted.[20]

Mourners and curiosity seekers clogged 40th and State Streets waiting to file in to see the battered body of their hometown boy. Many hailed from Mississippi and had escaped the violence of Jim Crow, which seemed to fall upon them now in Chicago. Thousands must have sent their children south in the summer to visit grandparents and cousins, dispensing the same lecture about the ways of white folks that Mamie had given Emmett. It could easily have been their child unspeakably slaughtered and sent home in a pine box. This realization inspired rage more than fear, because there on the South Side, surrounded by tens of thousands of African Americans, with a foothold, however tenuous, in the city's politics and media, they knew one thing for certain: they did not need to hide their anger anymore.[21]

When Mamie and her family arrived for the funeral service on Tuesday, escorted through the throng by half a dozen police officers, the barrel-vaulted sanctuary of Roberts Temple Church of God in Christ was full to its capacity of 1,500, and thousands more stood outside, where loudspeakers had been set up so people could hear what was happening in the church. Bishop Ford offered an emotional sermon based on the fiery text of Matthew 18:6. "But [whosoever] shall offend one of these little ones which believe in me, it were better for him that a millstone were hanged around his neck and that he were drowned in the depth of the sea."[22] Reverend Cornelius Adams asked the crowd to offer "fighting dollars" to help the struggle for racial justice; $700 was immediately taken up for legal aid.[23] Throughout the viewing and the funeral ushers collected money for the NAACP and Emmett's family, a process that would continue in the weeks and months ahead.

"I had no idea how I could make it through," Mamie recalled. "But I knew that I had to do it. And I knew that it wasn't going to get any easier as we prepared for what was ahead."[24] Now that she had the world's attention, she had to decide what to do with it. As she looked into the glass-enclosed coffin, she knew that a political and spiritual struggle lay ahead to make her son's death meaningful in ways that his life hadn't had time to be. In the face of this burden she began to lose her grip. The *Chicago Tribune* reported that she "collapsed and had to be assisted to a seat after she looked for the last time at her son's body."[25]

With fifty thousand coming on Friday night and forty thousand more on Saturday, followed by three more days of crowds lined up to view Emmett's body, it is hard to say how many people became witnesses in this way. The *Chicago Tribune* reported, "Capt. Albert Anderson, in charge of a large police detail at the church, said that more than 100,000 persons had viewed the remains of the youth." The *Chicago Defender*'s estimate was more than twice that number, 250,000: "All were shocked, some horrified and appalled. Many prayed, scores fainted, and practically all, men, women and children, wept."[26]

In Mississippi a few days later the front page of the *Greenwood Morning Star* reported that Reverend J. A. Perkins from Tupelo had declared at the National Baptist Convention in Chicago that the butchery of Emmett Till "did something to everybody." Before this tragedy opinion among African Americans about the path to full citizenship had been divided, he acknowledged. There would still be differences of opinion, but Till's slaying spurred and unified the fight against segregation. "We are not going to be afraid of anyone," Perkins vowed. "We are going to battle for what is right—as human beings—and we are going to stand against this wrong."[27]

David Jackson and Ernest Withers, photographers for *Jet* and *Ebony*, snapped pictures of Emmett's body at the funeral home, Withers a close-up and Jackson a full-body shot.[28] Several other magazines and newspapers printed photographs, but Withers's close-up of Emmett's face, published in *Jet* on September 15, four days before Roy Bryant and J. W. Milam went on trial for the killing, was passed around at barbershops, beauty parlors, college campuses, and black churches, reaching millions of people. Perhaps no photograph in history can lay claim to a comparable impact in black America.[29] "I think the picture in *Jet* magazine showing Emmett Till's mutilation was probably the greatest media product in the last forty or fifty years," Representative Charles Diggs said in 1987.[30]

Television coverage of the case had an even greater impact. Few in the summer and fall of 1955 could fathom the immense power of television. Even many sophisticated journalists and seasoned politicians did not yet understand that there was a new mass language that came of age with the Emmett Till generation. This was television's first real "media circus," and it made clear that civil rights stories would not be confined to a minority of Americans or a particular region. "The television cameras showed what the body looked like and showed the big crowds," recalled the journalist Harry Marsh. "All the networks that were operating at that time took film from the Chicago stations." It was television, Marsh continued, that "ignited the national interest in the story and that fed into the major coverage of the trial."[31]

The sociologist Adam Green observes that the spectacle surrounding Emmett Till's death "convened" black Chicago and black Mississippi into one congregation that trumpeted the tragedy to the world. These voices of mourning and protest emerged exactly as Mamie hoped they would. Members of this black national congregation launched rallies, letter campaigns, and fund drives that transformed another Southern horror story into a call for action. In this call and response, Green writes, "northern city and southern delta seemed the same place, and the need for collective action among African Americans across the nation seemed urgent as never before."[32]

9

WARRING REGIMENTS
OF MISSISSIPPI

In many ways Emmett Till was a casualty of the anger produced by the U.S. Supreme Court's decision in *Brown v. Board of Education*, handed down on May 17, 1954, first dubbed "Black Monday" by Representative John Bell Williams of Mississippi. Mississippi Circuit Court judge Thomas Brady speculated that the mandate to integrate public schools would compel right-minded white men to commit violence against foolhardy black boys. Killing would be necessary, even unavoidable. "If trouble is to come," Brady warned in his incendiary manifesto, *Black Monday*, "we can predict how it will start." The detonator would be the "supercilious, glib young Negro, who sojourned in Chicago or New York, and who considers the counsel of his elders archaic." That black child "will perform an obscene act, or make an obscene remark, or a vile overture or assault upon some white girl." This violation of segregation's most sacred taboo would set off a deluge of white violence against black boys. The foolish doctrine of equality between the races, cautioned Brady, was "the reasoning which produces riots, raping and revolutions."[1]

Brady's words are an almost eerie prediction of the murder of Emmett Till, but at the time there was nothing notable about his menacing fantasy of a race war. Such threats of violence were nothing new to black Mississippians or, for that matter, to any student of Reconstruction or the age of Jim Crow. But a new front in the war over Mississippi had opened after 1945, when African American soldiers returned from Europe and the Pacific. Into these renewed hostilities the Supreme Court dropped its *Brown v. Board of Education* bombshell. Milam and Bryant were not on a political mission when they pounded on Moses Wright's door, and they did not kidnap Emmett Till beneath the banner of states' rights, racial integrity, or white supremacy. The white men carried out their brutal errand in an atmosphere created by the Citizens' Councils, the Ku Klux Klan, and the mass of white public opinion, all of which demanded that African Americans remain the subservient mudsill of Mississippi—or die. But the battles ignited by *Brown* had been brewing for a long time.

According to William Bradford Huie, Milam later justified Till's lynching using the terms of violent racial and sexual politics:

> Just as long as I live and can do anything about it, niggers are going to stay in their place. Niggers ain't gonna vote where I live. If they did, they'd control the government. They ain't gonna go to school with my kids. And when a nigger even gets close to mentioning sex with a white woman, he's tired o' livin'. I'm likely to kill him. Me and my folks fought for this country, and we've got some rights. . . . "Chicago boy," I said, "I'm tired of 'em sending your kind down here to stir up trouble. God damn you, I'm going to make an example of you just so everybody can see how my folks stand."[2]

This kind of perverted logic, though Milam referred directly to *Brown*, had also motivated the bloody overthrow of Reconstruction in Mississippi in 1875, which was the last time the Magnolia State had outmatched the level of racial conflict in the state in the mid-1950s. It drove the violent massacre and coup in Wilmington, North Carolina, in 1898, which served as a model for the Atlanta "Race Riot" of 1906. It sparked massive mob violence against the black community in Springfield, Illinois, in 1908, which helped inspire W. E. B. Du Bois and his white and black allies to found the National Association for the Advancement of Colored People the following year. It set off the slaughter in Elaine, Arkansas, in 1919; and the fiery pogroms in East St. Louis in 1917, in which perhaps 200 blacks died; and Tulsa in 1922, in which as many as 300 blacks died and 10,000 were left homeless. The logic expressed by Judge Brady and J. W. Milam was all too familiar to nearly every American with dark skin.[3]

Like Judge Brady, J. J. Breland, one of Milam's attorneys and a Citizens' Council official, exported the blame for the lynching northward. Breland told Huie that Milam and Bryant "wouldn't have killed [Till] except for Black Monday. The Supreme Court is responsible for the murder of Emmett Till."[4] If that was indeed Milam's thought, it wasn't original to him, any more than it was to Breland. Southern white leaders liked to pretend that they were shocked, shocked, by the Court's heretical ambush. For anyone who had been paying attention, however, the Court's decision in *Brown* was not a surprise. White politicians howled as though Chief Justice Earl Warren had ordered the sneak attack on Pearl Harbor, but the armies of segregation had seen this attack coming. "Rather than a backlash against the unthinkable," writes the historian Jason Morgan Ward, "the segregationist movement was a coordinated assault against the foreseeable."[5]

The contours of the renewed battle had been visible before World War II. As early as 1941 Mississippi's former lieutenant governor E. D. Schneider defended the white primary and the poll tax at an African American agricultural fair in Clarksdale, promising that black people would never vote in Mississippi. The black crowd booed him loudly. Then a black newspaper editor stood and told the assembly that, in fact, black Mississippians would vote, and soon. And if necessary, he continued, blood would be shed in Mississippi as well as abroad to secure democracy.

The following year a member of the Yazoo–Mississippi Delta Levee Board warned, "We are going in the future to get quite close to the time when the darkey will be protected by Federal law in his vote in the South, and we all know what that will mean in Mississippi."[6] As the Magnolia State's own Richard Wright wrote that same year, "When one of us is born, he enters one of the warring regiments of the South."[7]

Mississippi emerged as the leader of the 1948 States' Rights Democratic Party, better known as the Dixiecrats. Led by Mississippi's governor Fielding Wright and South Carolina's Strom Thurmond, the Dixiecrats broke with President Harry Truman, the Democratic standard-bearer, over his support for civil rights legislation, an antilynching bill, measures to end the poll tax, and the establishment of a permanent Fair Employment Practices Commission. The Dixiecrat revolt of 1948 captured the electoral votes of Mississippi, Louisiana, Alabama, and South Carolina and presaged the rise of a two-party South.[8] Judge Brady was already a fuming Dixiecrat, calling for a new party "into whose ranks all true conservative Americans, Democrats and Republicans alike, will be welcomed" to battle "the radical elements of this country who call themselves liberals." Senator James Eastland of Mississippi termed the Dixiecrat revolt "the opening phases of a fight" for conservative principles and white supremacy, and "a movement that will never die."[9]

Despite its rhetorical confidence, the segregationist movement was haunted by a disquieting sense that it no longer possessed the political high ground. Walter Sillers Jr., a Delta cotton planter who served in the Mississippi State House from 1916 to 1966, wrote to Senator Eastland as early as 1950 that the African Americans were "well ahead of us" in the battle over civil rights: "They are holding meetings every week throughout the whole Delta counties and are well organized, and should an emergency arise they know exactly where to go and what to do, while our people are disorganized and would scatter like a covey of flushed quail."[10] If Sillers overestimated the level of preparation among black Mississippians, he was right about their intentions.

One of the key figures in the African American resistance to

Mississippi's racial caste system was a World War II veteran named Amzie Moore. Nothing about Moore's early life made it easy for him to imagine a world beyond white domination. He was born in 1912 on a cotton farm in rural Mississippi. His grandfather died at 104, dividing the land among his seven children, who promptly lost it all in the Great Depression. Amzie's mother passed away when he was still a boy, "so my father came and picked up the two smaller children and at about fourteen I was on my own." The abandoned youth lived hand to mouth, depending on various relatives and teachers, and was often hungry. He nevertheless graduated from high school in 1929, a rarity among his generation of African Americans in Mississippi. In 1935 the heavyset, round-shouldered Moore got a job as a janitor with the U.S. Postal Service and moved to nearby Cleveland, Mississippi. For a black man of his time this was an excellent job, and he kept it for more than thirty years. He also scrambled to make money as an entrepreneur.[11]

To Moore the color line in Mississippi seemed so stark and inexorable as to seem the very will of the Creator. "For a long time," he remembered, "I had the idea that a man with white skin was superior because it appeared to me that he had everything. And I figured if God would justify the white man having everything, that God had put him in a position to be the best."[12] He told an interviewer, "I just thought [whites] were good enough, that God loved them enough, to give them all these things that they had. And that, evidently, there had to be something wrong with me."[13]

Moore's deeply internalized sense of white supremacy never precluded his devoted hope that Mississippi blacks would one day rise to economic self-sufficiency and equal citizenship. When the freedom movement began to stir, Moore heard it as a call to arms. His "first knowledge of the freedom movement" came in 1940, when he joined several thousand African Americans to talk about agricultural modernization, school equality, and economic development. "Some 10,000 living in the Delta section in Mississippi gathered for a mass meeting," one of his colleagues recollected in a letter written twenty-five years later. "At that time we were pulling for the 'separate but equal' philosophy. Schools were separate but not equal."

Speakers came from Tuskegee Institute, Alcorn College, and Washington, D.C.[14] Moore remembered the meeting at Delta State College as "our first awakening" and "the beginning of the change."[15]

Like the African Americans who planned the meeting at Delta State, Moore defined racial uplift as entrepreneurial success: it wasn't just the nearest weapon to hand; it was among the most powerful. The year after the meeting Moore became the first African American around Cleveland to get a Federal Housing Authority loan: "I started out buying lots, building my first house in 1941, giving it bathroom facilities and everything there, gas, all the conveniences because I had to convince myself that I had the capability to do it."[16] He built his own brick house at 614 South Chrisman Street, in the middle of Cleveland's teeming "Low End" black business district. "Amzie," recalled Charles McLaurin, one of the many young civil rights workers Moore influenced, "was middle-class."[17]

World War II interrupted Moore's ascent in 1942, when he was drafted at age thirty. He was quickly introduced to the rude truth that a middle-class black man from Cleveland, Mississippi, was to the vast majority of his fellow Americans always first and foremost just a black man. "I really didn't know what segregation was like before I went into the Army," he recalled. At bases all over the country he experienced the segregation and mistreatment of black soldiers.[18] And he resisted with some success. Matthew Skidmore, who served with him in segregated units, including at Walterboro Army Airfield in South Carolina, wrote a letter to Moore in 1955 after he saw a photograph of his old comrade doing civil rights work in Mississippi: "Reading Ebony Magazine, I saw your picture and why it was there. Remember how we fought racial prejudice, especially in South Carolina? Remember how we won?"[19] In another letter his old friend reminded Moore of various episodes in their battles against white supremacy—"the theater incident" and "the busses." He wrote: "Remember we couldn't use the facilities of that Tropical Club on the base? Maybe you don't, but all that was corrected. Perhaps you don't remember, but the mayor promised that the white people in his little town would be courteous and respectful."[20]

What Moore remembered more clearly were the endless slights, insults, and even killings that occurred during his time in the service: "Everywhere we went, we were faced with this evil thing—segregation. If we were here fighting for the four freedoms that Roosevelt and Churchill talked about, then certainly we felt that the American soldier should be free first."[21] Unlike mustered white soldiers, black soldiers encountered the fight for democracy well before they shipped overseas. Moore found the ironies and idiocies of his racial predicaments nearly unendurable. "Here I am being shipped overseas, and I've been segregated from this man whom I might have to save or he save my life. I didn't fail to tell it."[22]

Ironically the military authorities selected Moore to sell his African American compatriots on their stake in the outcome of the war. Stationed with the Tenth Air Force in Myitkyina, Burma, Moore traveled throughout South Asia speaking about democracy and the war. "We had to counteract this Japanese propaganda by giving lectures to our soldiers. That was my job. We were promised that after the war was over, things would be different, that men would have a chance to be free. Somehow or other, some of us didn't believe it."[23] Whether or not Corporal Moore managed to convince any of the black troops, or even himself, that the war for democracy included them, too, his travels all over the United States and in Mexico, South Asia, China, Japan, and Egypt enlarged his perspective on the freedom struggle in Mississippi. "I think what God really did with me, in this particular thing, was put me on a ship and send me around the world. And let me live in different environments and be in contact with different people and to really and truly find out what was behind it."

When his unit was shipped from California to India, this descendant of slaves found himself walking one day beside the Great Temple in Calcutta. When he saw the ancient splendor and "people dying in the streets and people walking by them like they aren't even there," he believed that the wisdom that built that venerable civilization had "turned backwards" and the old regime had collapsed. "You can't set up an aristocracy and expect it to float," he ruminated. For every social order that goes up, Moore became convinced, "there's a coming down, and like the great wheel of time every spoke comes on over and then it goes down in the dirt." In that moment his mind turned to America—and no doubt to Mississippi. "We will have to get along with each other in this country because that's the only way you can survive another hundred or two or three hundred years."[24]

When the U.S. Army issued an honorable discharge to Corporal Amzie Moore on January 17, 1946, at Camp Shelby, Mississippi, he had a four-hour bus ride to consider how he would spend the rest of his days.[25] He planned to settle down and make a life for himself, but he was angry at the contradiction of fighting for human freedom with a Jim Crow army. He wondered whether there was any truth in what he had been instructed to tell the black troops about the postwar era.[26] When he arrived in Cleveland, he got his answer: whites there had organized a "home guard," ostensibly "to protect the [white] families against Negro soldiers returning home." Over the next few months a number of black veterans in the town were killed. "I think the purpose of the killing was to frighten other Negroes," Moore attested. "It certainly had its psychological effect."[27]

Moore remained determined to see his wartime propaganda about democracy move the country beyond empty rhetoric. With this in mind he set out to build a statewide network of African American activists to pursue the right to vote. Nothing else would be possible without it, he believed, and anyone who could count could see that black votes in the Delta would change things.[28] By late August 1955 Moore was a successful businessman running a gas station, beauty parlor, and grocery store. He was also the president of the Cleveland, Mississippi, branch of the NAACP, which had over five hundred members, and was an established force among black civil rights workers in the South.

According to Moore, it was shortly after J. W. Milam and Roy Bryant kidnapped Emmett Till that Moses Wright first called him. Like Reverend Wright, Moore's first response was hope: "I thought to myself, well, he's probably around Greenwood there somewhere. I never thought anybody was going to lynch him. [A few days later], I got another call." By that time the teenage fisherman had come across Emmett's body in the currents of the Tallahatchie. Now harder choices needed to be made.

Moore set out for Money even though friends told him, "'You'd better not go, they watching out for you, they're going to kill you.' But I went over there. And then when I got to Money nobody would tell me where Mr. Wright lived. He lived out from Money, but nobody would, they claimed they didn't know where he was. So I left and come back." This was far from the end of Moore's involvement in the case, however.[29] Over the previous twenty-four hours decisions had been made and events set on a course that would sweep up Mississippi and Chicago, and eventually the United States and the world.

The Till case affected everyone in Moore's activist network in Mississippi, many of whom were also World War II veterans unwilling to accept the old ways after a costly global crusade, ostensibly for universal democracy. They became what the historian John Dittmer calls "the shock troops of the modern civil rights movement."[30] Many became NAACP leaders, especially after the *Brown* decisions of 1954 and 1955 shone a spotlight on school desegregation and lifted up new leaders who rejected the "separate but equal" framework. Moore's network included Medgar Evers; E. J. Stringer, a Columbus dentist who would be elected president of the state conference of NAACP branches in 1954; Aaron Henry, a druggist from Clarksdale who would become an important civil rights leader; the voting rights advocates Reverend George Lee and Gus Courts; and Dr. T. R. M. Howard, a charismatic and daring physician from Mound Bayou.[31]

Few were more involved than the intelligent, serious-minded Medgar Evers. His military service in the thick of the European theater during World War II had given him gravity beyond his years. "He went to the army and fought for this country and came out and we wasn't enjoying the freedom we fought for," his brother Charles observed. For Medgar voting was the central litmus test of democracy, especially in Mississippi. "Our only hope is to control the vote," he said.

In 1946 Medgar and Charles and four friends registered to vote at the courthouse in Decatur, Mississippi. When they returned to cast their ballots in the Democratic primary on July 2, however, a small white mob blocked the courthouse steps. "When we got into [the clerk's] office, some 15 or 20 white men surged in behind us," Medgar recalled. Menaced with guns as they were leaving, they went home, got their own guns, and returned to the courthouse. Leaving their weapons in the car, the Evers brothers and their companions tried to walk in again, but once more the way was blocked. "We stood on the courthouse steps, eyeballing each other," Charles recalled. Finally Medgar decided to avoid what seemed inevitable bloodshed and said, "Let's go, we'll get them next time." "More than any other single thing," Charles claimed, "that day in Decatur made Medgar and me civil rights activists."[32]

Medgar's first order of business was to use his GI Bill benefits to attend college. He enrolled at Alcorn State College that fall of 1946. Muscular and athletic, he played running back on Alcorn's football team. He also served as editor of the *Greater Alcorn Herald* and was elected president of the junior class. Evers worked hard, made good grades, labored over his vocabulary, and became a faithful reader of newspapers. He met and married Myrlie Beasley from Vicksburg, a brilliant and beautiful young woman who would become a full partner in his civil rights work. According to a friend, he also joined a monthly interracial discussion group on world affairs at all-white Millsaps College. He developed a keen admiration for Jomo Kenyatta, leader of the Kenyan anticolonial struggle against the British; in fact, in 1953 Medgar and Myrlie christened their son Darrell Kenyatta Evers. Kenyatta led an armed revolution, and Medgar's admiration suggests that he was willing to consider any means necessary to overthrow white supremacy in Mississippi. For African Americans in the South the means of effecting change were always limited and fraught with risks, but most were at least considered, their consequences carefully weighed. The historian Charles Payne writes that Medgar "thought long and hard about the idea of Negroes engaging in guerrilla warfare in the Delta" but ultimately could not square it with his religious beliefs.[33]

Soon after he graduated from Alcorn, Medgar renewed an acquaintance with T. R. M. Howard and became a sales representative for

Howard's Magnolia Mutual Life Insurance Company. He and his bride moved to Mound Bayou, the historic all-black community in the Delta. His new job required he drive the lonely roads of the Delta, and he became sharply aware of the harsh lives of Mississippi's black sharecropping families. "He saw whole families there picking cotton, living like slaves," Charles recounted. The Evers brothers came to believe that black Mississippians' suffering was the result of economic exploitation, the threat of physical violence from whites, and the denial of their right to the ballot. "Medgar vowed to improve these people's lives," said his brother.[34]

Theodore Roosevelt Mason Howard was the tall, light-skinned head of surgery at Mound Bayou's Taborian Hospital, whose all-black staff served the black community. No civil rights advocate in the state of Mississippi could match Howard's charisma, charm, and wealth. In addition to his plantation he owned a housing construction firm, a beer garden and restaurant, some livestock, and an insurance firm. He could afford servants and chauffeurs; pheasants, quail, and hunting dogs; and a fleet of fine automobiles, as well as a Thompson submachine gun and a number of other weapons. These were simple acknowledgments of his having achieved stature and influence in a society that demanded he never wield it. "One look," wrote Myrlie Evers, "told you he was a leader: kind, affluent, and intelligent, that rare Negro in Mississippi who had somehow beaten the system."[35]

Howard recruited Amzie Moore and a few others to go into business with him. According to Moore's canceled checks, he invested $1,000 in the Magnolia Mutual Life Insurance Company in 1951, when Howard founded the company. Within a couple of months Moore was appointed to Magnolia Mutual's board of directors, along with Aaron Henry. Soon Medgar Evers joined them.[36] As ever among Mississippi's black leaders, economics and politics were not far apart.

On the morning of December 28, 1951, Howard, Moore, and Henry, among others, founded the Delta Council of Negro Leadership, soon renamed the Regional Council of Negro Leadership (RCNL), "to guide our people in their civic responsibilities regarding education, registration and voting, law enforcement, tax-paying, the preservation of property, the value of saving, and to guide us in all things which will make us stable, qualified, conscientious citizens, which will lead to first class citizenship for Negroes in the Mississippi Delta and the State of Mississippi." Moore became the first vice chairman and Howard the founding president. Henry called the RCNL a kind of "homegrown NAACP."[37]

The RCNL's focus on what Moore called "the changing of the economic standpoint," however, differentiated it from the NAACP, which downplayed economic development in favor of civil rights.[38] The RCNL billed itself as an organization "for the common good of all citizens in the Delta area and state, regardless of race or creed."[39] It was trying to present to its white neighbors a more acceptable face than the "radical" NAACP.[40] Though in the early days of the RCNL economic development was the first concern, civic questions—voting, police brutality, and the indignities of segregation—quickly emerged. Moore later claimed the RCNL ultimately had 100,000 members from forty counties in Mississippi alone; these included numerous professional people, "principals of schools, people of the Department of Agriculture, teachers."[41]

Evers soon became the RCNL's program director, playing a key role in the 1952 campaign targeting restroom facilities in service stations. African Americans in Mississippi knew the dangers and indignities that confronted blacks on the highways of the South. Most white-owned establishments along the highway marked their public restrooms "White only" and offered no facilities for blacks. So African American women, children, and men had to take to the roadside woods and thickets to relieve themselves or wait uncomfortably until they got where they were going. If that were not enough, Mississippi state troopers were well known to treat black motorists with a lack of courtesy that often extended to outright brutality. Therefore the first crusade of the RCNL was for equality on the public roads and featured the slogan "Don't Buy Gas Where You Can't Use the Restroom."[42] Evers and Howard paid a black printer in Jackson to produce tens of thousands of bumper stickers emblazoned with the phrase, and, according to his brother, Evers "gave out that bumper sticker to hundreds of Negroes statewide—whoever would take one."[43] "As time went by," Myrlie wrote later, "Medgar and I would see little bumper stickers with those words on the usually beat up automobiles of Delta Negroes." Here was a portent of all that was to come.[44]

The RCNL's crusade mobilized large numbers of African Americans, who declined to buy gasoline from service stations that offered them no facilities. Though technically the protest fit neatly inside the "separate but equal" system of Southern segregation, African Americans' self-assertion violated the social arrangements that segregation was fashioned to teach and defend. Victory represented a crack in the old order: most white-owned service stations soon chose to provide segregated restrooms, and many even posted signs announcing "Clean Rest Room for Colored."[45]

As impressive as these victories must have seemed to blacks in the Delta in the early 1950s, the RCNL's stupendous rallies in Mound Bayou were its trademark. Under a massive circus tent on Howard's plantation, thousands of African Americans gathered to hear national speakers and enjoy renowned musical artists. Literally tons of ribs and chicken were served. At the RCNL's first annual conference in 1952, Representative William Dawson from Chicago became the first black congressman to speak in Mississippi since Reconstruction, addressing a crowd of seven thousand. The great gospel singer Mahalia Jackson appeared alongside him.[46] At the RCNL's third annual meeting, in 1954, only ten days before his historic victory in *Brown*, Thurgood Marshall spoke to roughly eight thousand attendees, accompanied by eight school bands and the sixty-piece marching band from Tennessee A&I State University, which led the "Great Freedom Parade" down Main Street while Howard and Marshall waved from a convertible. That evening's panel discussion in the big tent was "The Negro in an Integrated Society."[47] In 1955 *Ebony* featured a photograph of the circus tent with thirteen thousand assembled beneath it to hear the newly elected black congressional representative Charles Diggs from Detroit.[48]

At each of the annual rallies the RCNL's Committee on Voting and

Registration held workshops as part of its drive to educate and register new black voters. In many areas of Mississippi sheriffs refused to accept poll taxes from African Americans, effectively barring them from registering to vote. Courthouse clerks turned others away for no reason more compelling than their own objection to black voting. For those rare places where voting tests were reasonably administered, the RCNL conducted classes on the Mississippi state constitution so that aspiring black voters might pass. The number of registered African Americans slowly increased to a twentieth-century record of twenty-two thousand in 1954.[49]

The RCNL's voter registration campaign took place against a growing white resolve to protect "the Southern way of life" from black self-assertion and federal intrusion. Even though their economic system relied at least as much on government subsidies as it did the world cotton market, Mississippi planters feared that the river of federal money would rise with strings attached. From the very instant that African Americans in the North entered FDR's New Deal coalition, white supremacist leaders preached gloom and doom for Mississippi race relations. Persuaded that FDR's coalition of liberals, African Americans, and labor unions imperiled "the white democracy of the South," white diehards in Dixie launched a movement to defend segregation and preserve their political power, whether that meant a bloody battle for control of the Democratic Party or abandonment of the party of their fathers altogether. "A few far-sighted Southern Democrats realized in 1936 that the great Democratic River was beginning to split into two forks," wrote Judge Brady in 1948 as he joined the Dixiecrat armies.

Education and voting were divisive years before the *Brown* decision. In 1942 the editor of the *Meridian Star* rejected the idea of federal aid for education because "camouflaged education help" would bring Mississippi "one step closer to the intertwined evils of Hitlerized, totalitarian rule and social equality." Two years later a former governor of Mississippi, Mike Connor, lashed out at the *Smith v. Allwright* decision, which struck down the "white primary." New Dealers, said Connor, had embraced "un-American and undemocratic philosophies of government" in an effort "to change the very form of government from a republic to an absolute, totalitarian state of communism or national socialism, which would destroy at home everything our armed forces abroad are fighting to preserve." The national Democratic Party had fallen so low that it would "traffic with northern Negroes to place the black heel of Negro domination on our necks." During the 1946 election the notorious senator Theodore Bilbo explained why it was crucial to stop blacks from voting, citing folk wisdom that would endure among whites well into the 1960s: "If you let a few register and vote this year, next year there will be twice as many, and the first thing you know the whole thing will be out of hand." Bilbo offered to give county registrars, who administered the voting tests, "at least a hundred questions no nigger can answer." For good measure he added his favorite admonition: "The proper time to manage the nigger vote is the night before the election."[50]

And the warring regiments of Mississippi persisted in this slow trench warfare for many years before nine white men in black judicial robes dropped the bomb on public school segregation.

10

BLACK MONDAY

On May 21, 1954, four days after the U.S. Supreme Court's ruling in *Brown v. Board of Education*, Judge Brady wrote the second of two fervent letters to Representative Walter Sillers Jr. urging a political response. He envisioned a new political formation that would "cut across all factors, political groups, and embody leaders in every clique." He continued, "All white men in every walk of life must be mustered out. It must be made their fight. If the Southern states do not unify in thought and action, the NAACP will emerge victorious."[1]

Brady had attended New Jersey's Lawrenceville preparatory academy and graduated from Yale University in 1927. He took his JD in 1930 from the University of Mississippi, where he taught sociology for two years, and was vice president of the state bar association.[2] Ever since, he had practiced law and then donned judge's robes in the small town of Brookhaven. In time he would serve on the Mississippi Supreme Court.[3]

A week after writing his letters to Sillers, Brady took the podium at a meeting of the Sons of the American Revolution in Greenwood, Mississippi, and delivered a fevered speech entitled "Black Monday." Its central themes were the legal fallacies and political perils in *Brown*. The Court's decision rested upon "Communistic" sociological arguments and the Fourteenth Amendment, he argued. The former were pernicious and irrelevant, but the amendment's assurance of federal citizenship guaranteed in its equal protection and due process clauses "was filled with dynamite." Federal troops and carpetbagger legislatures had imposed the unlawful Fourteenth Amendment after the War Between the States, but it had "never been of any moral force in the South." The illegitimate ruling in *Brown* ignored more than half a century of legal precedent. The NAACP and the international communist conspiracy were the driving political forces behind the decision.[4] Brady recalled that after the speech "several men came up and said, 'Judge, you ought to write that in a book.' I told several men in public office that I was going to wait until June and if nothing was done about the problem, I was going to publish it. Nothing was done, so I put it out."[5]

Brady stretched his oration to ninety-seven pages by adding a generous slathering of white supremacist clichés disguised as scientific facts and paranoiac ravings dressed up as patriotic tub-thumping. Like most similar fare, its driving engine was the gnawing terror of "miscegenation" between black men and white women. The speech and its pumped-up sequel inspired the founding of the Associated Citizens' Councils, which published the speech in July as a booklet titled *Black Monday: Segregation or Amalgamation. America Has Its Choice*.[6] Florence Mars, a Mississippi native who attended the trial of Emmett Till's murderers, called it "a remarkable document, not only because of its large impact in the fight against complying with the decision but also because [among white Mississippians] its views on race were widely accepted at face value for years."[7] Its circulation throughout the Deep South arguably made it the most influential political statement of the white South in the wake of *Brown*, particularly since it became the founding handbook and doctrinal scripture of the Citizens' Council movement.[8]

The imagined beastly lust of black men for white women seized Brady's pornographic political imagination, and the unsullied Southern white woman became the most important symbol of white male superiority.[9] "The loveliest and purest of God's creatures, the nearest thing to an angelic being that treads this celestial ball is a well-bred, cultured Southern white woman or her blue-eyed, golden-haired little girl," he wrote. "The maintenance of peaceful and harmonious relationships, which have been conducive to the well-being of both the White and Negro races of the South, has been possible because of the inviolability of Southern womanhood."[10]

To that very day, insisted Brady, the hope of black citizens lay in an America ruled by an unchallenged white supremacy. "The date that the Dutch ship landed on the sandy beach of Jamestown was the greatest day in the history of the American negro."[11] Their white benefactors forced the African slaves "to lay aside cannibalism" and their other "barbaric savage customs." The enslavers gave the enslaved a language, moral standards, and a chance at Christian salvation, although the African American had never absorbed any of them fully enough to transcend his primitive nature. "The veneer has been rubbed on, but the inside is fundamentally the same. His culture is superficial and acquired, not substantial and innate."[12] Brady declared, "You can dress a chimpanzee, housebreak him, and teach him to use a knife and fork, but it will take countless generations of evolutionary development if ever before you can convince him that a caterpillar or a cockroach is not a delicacy." The chimpanzees were not alone in their uncivilized shortcomings. "Likewise the social, political, economic and religious preferences of the negro remain close to the caterpillar and the cockroach."[13]

The evils of "amalgamation" with this inferior race justified extreme authoritarian measures in defense of white supremacy. Brady's prescription included the abolition of public schools, if necessary; the intimidation of rebellious African Americans by economic reprisals; and the institution of special courts to try and punish "all undesirables, perjurers, subversives, saboteurs and traitors."[14] Not to fight the Supreme Court's usurpation of the Constitution "is morally wrong and the man who fails to condemn it and do all that he can to see that it is reversed is not a patriotic American."[15] It was not the first or last time an elected Southern official in the grip of racist outrage painted breaking the law as patriotism, but the Cold War gave Brady a new set of brushes. He cast the broader cause as "stopping and destroying the Communist and Socialist movements in this country." Race agitation was a tool of the "Red Conspiracy," a means to an end "in the overall effort to socialize and Communize our Government."[16]

Some years afterward Robert "Tut" Patterson, at thirty-two the first executive secretary of the Citizens' Council, told Judge Brady that his decision to devote the rest of his days to fighting desegregation came after he read *Black Monday*. A cotton planter and cattle rancher from Holly Ridge, Mississippi, Patterson penned a protest letter that helped inspire the founding of the Citizens' Council. "I, for one, would gladly lay down my life to prevent mongrelization," he wrote. "There is no greater cause." He felt that those who shared his point of view faced formidable adversaries but would triumph in the end. "If every Southerner who feels as I do, and they are in the vast majority, will make this vow, we will defeat this Communistic disease that is being thrust upon us."[17]

The author of *Black Monday* had watched integration coming for a long time, but the menace grew closer with every passing year. On June 5, 1950, the U.S. Supreme Court ruled in *Sweatt v. Painter* that the University of Texas Law School had to admit African Americans because no separate law school the state might build would equal the state university's law school in prestige and opportunity. On the same day the Court ruled in *McLaurin v. Oklahoma State Regents* that the University of Oklahoma Law School could not simply segregate black students inside the same law school.[18] Representative William Colmer of Mississippi replied, "It means there will be an ever increasing intermingling of Negroes and whites in public places. It is the forerunner of a final decision by the Court at a not too distant date, denying segregation in all public institutions, both State and Federal."[19]

Mississippi's white political elite had anticipated the *Brown* decision in 1954. That spring the state legislature put forward an "equalization" program for teachers' salaries, bus transportation, and school buildings but delayed the program's launch until after a favorable ruling from the Supreme Court; any ruling along the lines of *Brown* would kill it. The school budget normally accounted for a two-year period, but in 1954–55 the immediate funding applied for only that year; the way forward would become clear with the Court's decision, they reasoned. At this late date, in other words, they offered to comply with the 1896 "separate but equal" ruling in *Plessy v. Ferguson* if the Court would be so kind as to avoid ruling against them in *Brown*. That they could in no way afford to equalize black and white schools without an unthinkable tax hike was beside the point.[20]

Well over six feet tall with a sheaf of red hair, Tut Patterson was a former Mississippi State football star and proud of his service in World War II. After the war he managed his family's Delta cotton plantation, but eventually he focused his energies on "states' rights and racial integrity," soon to be the banner of the Citizens' Council.[21] In early July 1954, Patterson met with David H. Hawkins, the manager of Indianola's cotton compress; Herman Moore, the president of a bank there; and Arthur B. Clark Jr., a local attorney with a degree from Harvard Law School. On July 11, two months after the *Brown* decision, these men met with fourteen of Indianola's civic and business leaders, including the mayor and the city attorney, at Hawkins's home. "We elected the banker president. I was the secretary," Patterson recalled. "And in the space of a few short months, it spread. The organization spread all over the Deep South and into other states." The manager, banker, and lawyer set it all in motion with a late-July town hall meeting of nearly a hundred people that founded the first Citizens' Council.[22]

Judge Brady was the chief attraction at the Indianola town hall, where he delivered another version of his "Black Monday" oration. He first framed the new organization as a legal public group, respectable and law-abiding, that would not encourage or participate in violence. "None of you men look like Ku Kluxers to me," he told them. "I wouldn't join a Ku Klux—didn't join it—because they hid their faces; because they did things that you and I wouldn't approve of."[23] But thereafter he fed the crowd on the raw meat of sex and race. "School integration is the first step toward racial intermarriage," he warned. "Wherever white men infused their blood with the Negroes, white intellect and white culture perished. It happened tragically in Egypt, Babylon, Greece, Rome, India, Spain and Portugal. When the NAACP petitioned the Court for integration, it was to open the bedroom of white women to Negro men."[24]

"I joined the Citizens' Council," said one Delta physician. "They would come up with all of this stuff about how the black boys might molest the white girls—that was a fear. Of course, one of the fears expressed by people like Jim Eastland, Ross Barnett, Judge Brady, and the man down in Louisiana, Leander Perez [was that] they always invoked the fear of intermarriage." And so these reasonable, respectable white men who disavowed Ku Kluxers stewed in their politicized racial fears until they became comfortable, tacitly or directly, with horrors. The doctor continued, "People like Andrew Gainey over in Meridian would say, 'Well, we can go to deportation, we can go to amalgamation, or we can go to extermination.' "[25]

Brady's screed reflected the spirit of panic that prevailed in white Mississippi and drove the growth of the Citizens' Councils. "The main way Councils were organized was through the service clubs," said William Simmons, the leader of the Jackson Citizens' Council and the son of a prominent financier. "Patterson and I would go and make a talk to Rotary or Kiwanis or Civitans or Exchange or Lions. We'd tell them what the Council movement is, what fellows were doing in different communities. Invariably the response was favorable." Simmons and Patterson recruited dozens of other speakers for the Council cause, and they fanned out all over the state.[26] Membership claims and estimates vary somewhat, but roughly nineteen members in July 1954 grew to twenty-five thousand in October, when Patterson established the first statewide office of the Association of Citizens' Councils of Mississippi in Winona, though it soon moved to Greenwood. A year after the first meetings in Indianola the Citizens' Councils boasted sixty thousand members in 253 communities and seven states.[27] By eighteen months out, Simmons claimed that the Councils had more than half a million members; independent investigators suggested that the truth was at least 300,000.[28] Patterson was astonished, he said, that the gathering of men who vowed to protect white womanhood would "expand miraculously into a virile and potent organization."[29]

The feral heart of Citizens' Council ideology drew almost equally on what W. J. Cash had fifteen years earlier termed "the Southern rape complex" and the delusion that international communism had spawned the civil rights movement.[30] On Black Monday, the Mississippi Association of Citizens' Councils charged, the U.S. Supreme Court "based their decision upon the writings of communists and socialists."[31] The NAACP was "a left-wing, power-mad organ of destruction" that had been "infiltrated by communist sympathizers." This language presented anew the logic on which American racism has long relied. Like many white citizens over the years, members of the Council believed that anything that weakened white supremacy or challenged the existing social hierarchy in any way was socialism. But this was largely code for preserving the country's racial caste system, centuries in the making. Animated by this fear, the Citizens' Council pledged to defeat integration and deliver "a complete reversal of the contrived trend toward a raceless, classless society."[32]

The Southern writer Lillian Smith captured the Citizens' Councils clearly and succinctly: "Some of these men are bankers, doctors, lawyers, engineers, newspaper editors, and publishers; a few are preachers; some are powerful industrialists. It is a quiet, well-bred mob. Its members speak in cultivated voices, have courteous manners, some have university degrees, and a few wear Brooks Brothers suits. They are a mob, nevertheless. For they not only protect the rabble, and tolerate its violence, they think in the same primitive mode, they share the same irrational anxieties, they are just as lawless in their own quiet way, and they are dominated by the same 'holy ideal' of white supremacy."[33]

Representative Wilma Sledge announced the birth of the Citizens'

Council movement from the floor of the Mississippi legislature, adding, "It is not the intent or purpose of the Citizens' Councils to be used as a political machine." They made haste, however, to claim credit for the passage of two constitutional amendments adopted in late 1954 and began to throw their weight behind politicians who backed their program.[34] The Councils also gained the early support of Senator Eastland, scores of elected officials, and the Hederman family, which controlled the *Jackson Daily News* and the *Clarion-Ledger*.[35] Their membership rolls soon included governors, legislators, and mayors and virtually all who aspired to those offices. In only a few months the Citizens' Council had become a huge organization that could speak for nearly all vested authority in Mississippi. "The Councils were eminently respectable," writes the historian Charles Payne, "and in Mississippi were hard to distinguish from the state government."[36] In fact, the state government eventually provided a good deal of funding for them. By 1955 the *Jackson* (MS) *Daily News* was printing Citizens' Council press releases as though they were reported news. Newspaper advertisements for Council membership gave no address or telephone number for those wishing to join but instead directed the reader to inquire at "your local bank."[37]

Some of the Council's power came from its sophisticated communications apparatus. This quickly mutated from a single duplicating machine to a propaganda mill spanning radio, television, and a phalanx of speakers for any occasion. It soon included legislators, governors, U.S. senators, mayors, and virtually anyone who aspired to electoral office.[38] Much of this noise machine focused on correcting supposedly unfair or inaccurate views of the South in general and Mississippi in particular. But the river of mail that poured out of Council offices was largely race-hate literature printed by Ellet Lawrence, a man whose unmatched power in the organization came from his ownership of a Jackson printing company. Titles included such "educational material" as "The Ugly Truth about the NAACP, Mixed Schools and Mixed Blood."[39] The mainstay of Council literature remained sexually provocative photographs of black men and white women drinking, dancing, or embracing, accompanied by breathless rants against race-mixing.[40]

This incendiary propaganda was taken as gospel truth by much of the white South. This helps explain the response of Council members to the scorn the world heaped on Mississippi in the wake of the lynching of Emmett Till. Outside condemnation enraged white Mississippians, most of whom saw the press reports as grossly unfair. To resist these slanders against the Magnolia State, the Citizens' Council leadership launched its own monthly newspaper, the *Citizen*, in the eleven states of the former Confederacy plus Missouri.[41] William Simmons became the first editor and wrote most of the stories for what soon became forty thousand subscribers. The paper illuminated the views and activities of the Councils and portrayed black Southerners as loyal darkies, utter buffoons, or, as one observer put it, "the Mau Mau in Africa."[42]

Economic reprisals against anyone, black or white, who favored racial equality were the Councils' standard method. Black teachers who overstepped Jim Crow's boundaries, openly favored racial equality, or were known to have joined the NAACP could count on losing their job. Black sharecroppers who registered to vote, signed a school desegregation petition, attended an NAACP meeting, or otherwise made known their dissatisfaction with the prevailing social order were in for trouble. "We won't gin their cotton; we won't allow them credit; and we'll move them out of their houses if necessary, to keep them in line," said one Yazoo County planter.[43]

Council members would obtain the names of dissidents and then contact their employers. If the employers were sympathetic, the member would ask them to tell the offending employees to take a vacation. If they ceased the offending activity, the employer might let them come back to work. If not, the "vacation" was permanent. In cases where the intransigent citizen was an independent merchant, farmer, or craftsman, credit could be cut off or wholesalers persuaded to stop providing necessary supplies. If these means did not prove effective, a visit from a local Citizens' Council member was in order. "They'd come and tell them, 'You've lived in this community a long time and if you want to stay here in peace, you'd better get your name off this list,'" explained Medgar Evers. Personal visits tended to be persuasive as the implied threats grew more and more clear. The Citizens' Councils could publicly declare their unwillingness to use violence because other whites, including Council members, were reliably willing to exercise it, and African Americans, without meaningful benefit of laws or police, were grossly exposed. The NAACP did what little it could. "We are investigating every case of intimidation that comes to our attention and action is being taken as fast as we can move," reported Ruby Hurley of the NAACP's Southeast Regional Office in September 1955. "We have conferred with officials in the Department of Justice about the murders and other acts of violence and intimidation which have occurred in Mississippi."[44]

The state NAACP president E. J. Stringer reported all manner of pressures and reprisals against him in the wake of the 1954 school petitions. His dental supplies vendor suddenly refused him credit. His insurance company canceled his policies. Loyal patients told him they had to patronize other dentists or risk losing their job. The IRS audited his finances. Banks in Mississippi refused to loan him money. His wife lost her teaching job. Death threats made answering the telephone a dicey proposition. The Stringers began to sleep in a middle bedroom as a precaution against bombings. "I had weapons in my house," he recalled. "And not only in my house, I had weapons on me when I went to my office, because I knew people were out to get me." He kept a pistol out of sight but close at hand even when he worked on teeth or did his paperwork. "I would take my revolver with me and put it in the drawer, right where I worked." Brave soul that he was, Stringer decided not to run for reelection in late 1954, and Dr. A. H. McCoy, a physician from Jackson, replaced him.[45]

McCoy was barely in office before the death threats began. The *Jackson Daily News* warned that if the NAACP persisted in pushing school integration petitions, "bloodshed" would be inevitable. After the NAACP filed its local school petition in Jackson, Ellis Wright, president of the local Citizens' Council, snapped, "We now tell the NAACP people they have started something they will never finish." McCoy took the U.S. Supreme Court at its word, however. He replied, "We just gave them the courtesy title of petitions. They were more in the nature of ultimatums." If violence erupted, he assured the newspaper's readers, black blood would not run down the streets alone: "Some white blood will flow, too."[46]

The editors of the *Jackson Daily News* warned in a front-page editorial that McCoy had crossed a line with his defiant remarks and that "self-respecting, law-abiding, peace-loving, hardworking Negroes" would not "follow [McCoy] in an attempt to force Negro children into white schools." They could not possibly want these radical measures. It was therefore up to this respectable class of African Americans to "openly repudiate McCoy, put a padlock in his mouth and [make] a summary end to his activities that will, if left unchecked, inevitably lead to bloodshed. . . . If not suppressed by his own race, he will become a white man's problem."[47]

The Citizens' Councils primly disavowed the violence and the kind of threats that had worn down Stringer. Judge Brady even claimed that the Councils "were the deterrent, the one deterrent, that kept [out] the organization of mobs and the operation of lynch laws in Mississippi."[48] Their day-to-day language, however, was the language of battle. Their flyers openly reflected intimidation and threats. The national office of the NAACP granted that perhaps the Citizens' Council did not itself order or commit acts of terrorism, violence, and murder, but it did create "the atmosphere in which it was possible for the Chicago boy, Emmett Till, to be murdered and for the perpetrators of the crime to escape justice." The power of the Council surely encouraged terrorists to believe they could act without fear of punishment.[49]

Tut Patterson knew that white terrorism was endemic to the segregationist cause but dismissed occasional violence as the irresponsible errors of "certain crackpots, fanatics and misguided patriots," for which the Citizens' Council movement could hardly be blamed.[50] In fact, the Councils claimed, their organization actually existed in large measure to avoid the clashes that the NAACP's extremism made almost inevitable. Despite the official line, however, violence often followed when the local Council's efforts to coerce or intimidate civil rights advocates did not have the desired effect. The Council's relationship to violence went deeper than that, however. One of the founders and financiers of the Citizens' Council movement said of the Till murder that it was "a shame that [Milam and Bryant] hadn't slit open [Till's] stomach so that his body would not have risen in the river."[51] Eight years after the murder, when a Council member murdered NAACP field secretary Medgar Evers, three bank presidents in Greenwood headed the White Citizens Legal Fund that financed the defense of the killer. Stated the Council, "We do not condone the murder of Medgar Evers and, of course, we have no idea of the guilt or innocence of the accused but we feel he is entitled to a fair trial."[52] This reciprocal arrangement was transparent. When one small town debated swimming pool desegregation in the mid-1950s, a patrician Citizens' Council member suggested, "I figure any time one of them gets near the pool, we can let some redneck take care of him for us."[53]

Phillip Abbott Luce, a white scholar who infiltrated the Citizens' Council in Mississippi soon after the lynching of Emmett Till, reported that he'd been told "the Council can do anything the Klan can do if it has to."[54] Only a few days before Till's murder, Ruby Hurley wrote to Gloster Current, the NAACP's national director of branches, to inform him that in the Mississippi Delta "the situation is at the point of explosion, particularly in the western part of the state, which includes the Delta." Hurley cited the "inflammatory editorials and columns" in the *Jackson Daily News*. "The White Citizens' Councils are becoming more brazen in their intimidation tactics—telephone calls, sending of hearses as was done to Jasper Mims of our Yazoo City Branch, threatening letters through the mails, circulation of the scurrilous literature enclosed etc. etc." She noted, too, the popularity of Judge Brady's *Black Monday* and the murder of a voting rights activist on the Brookhaven Courthouse lawn. "Reports from Cleveland"—Amzie Moore's hometown—"are bad." One of the FBI agents sent to investigate a murder in Belzoni was a close friend of Citizens' Council officials in a nearby county, she lamented. The last sentence in her report, filed only days before the murder of Emmett Till, was this: "Something must be done to protect our people."[55]

• • •

Despite the terror of the mid-1950s, both the RCNL and the NAACP in Mississippi had some cause for modest optimism: the RCNL's campaigns had won small victories, and their rallies attracted thousands, and NAACP membership grew steadily. Several thousand African Americans added their names to the voting rolls, swelling their number to the highest tally since the violent overthrow of Reconstruction eighty years earlier. Lawsuits and the NAACP's long march through the nation's courts persuaded two Mississippi governors in succession to begin to equalize teachers' salaries and build new black schools. In 1953, a year before *Brown*, the state senate refused to pass a proposed constitutional amendment that would have allowed the legislature to abolish public schools should the U.S. Supreme Court mandate school desegregation. In this volatile atmosphere *Brown* emboldened the Mississippi NAACP to strike hard at school segregation and to demand voting rights. The white response would plunge Mississippi into violent racial battles unmatched since the bloody 1870s.

In the wake of *Brown*, the state's NAACP president E. J. Stringer decided to follow the national NAACP's agenda of seeking "the removal of all racial segregation in public education . . . without compromise of principle." It was one thing to pass such resolutions in Atlanta or to approve of them in New York, but quite another to act on them in Mississippi. Nonetheless Stringer pushed the state branches to petition their local school boards to "take immediate steps to reorganize the public schools . . . in accordance with the constitutional principles enunciated by the Supreme Court on May 17" and to threaten legal action if necessary.[56]

The uncertain legal power of these petitions and the "communist-inspired" NAACP fueled paranoia if not panic among many white Mississippians. In Amite County a mob of twenty or so whites led by the sheriff and a member of the school board burst into an NAACP meeting at the local black school, snatched away the chapter's records, and interrogated the members. In Kemper County a large mob of heavily armed whites led by the sheriff appeared on the first day of classes at an all-white school because of alarming rumors that a group of black parents intended to enroll their children. In the fall of 1954 the Mississippi state legislature resubmitted a constitutional amendment to give them legal power to abolish the public schools and replace them with publicly funded "private" academies.

That summer thirty African Americans in Walthall County petitioned to allow their children to attend the previously all-white public schools there. School officials responded by closing the local black schools for two weeks and firing a school bus driver who had signed the petition.[57] At the subsequent hearing the board chair exclaimed, "Nigger, don't you want to take your name off this petition that says you want to go to school with white children?"[58] After further protests and threats from local whites, all thirty of the petitioners withdrew their names. Some insisted they had not understood the petition's contents and thought it was seeking equal school facilities, not attendance at white schools.[59]

It is doubtful that the petitioners in Walthall County actually misunderstood their own desegregation petition. It is certain that they knew their pursuit of justice put their children directly in harm's way, which helps explain why some African American leaders in the state declared equal facilities a sufficient ambition. Unable to reach any lasting consensus among themselves, many thought school equalization—the fulfillment of *Plessy*, not *Brown*—seemed both more likely and perhaps more desirable. J. H. White, president of Mississippi Vocational College, pleaded with the NAACP leadership to pursue "adequate facilities and other things that our people need first and when you lay that foundation you have made a great contribution and many other problems will be solved in years to come."[60] Neither side, White believed, was ready for full-blown integration. Far more often black leaders simply agreed with T. R. M. Howard's astute assessment that "to petition school boards in Mississippi at the present time is like going to hunt a bear with a cap pistol."[61]

That the NAACP in Mississippi proceeded to battle for school desegregation and voting rights in this atmosphere is remarkable. Six days after the U.S. Supreme Court issued its decision in *Brown II*, telling school authorities to move "with all deliberate speed," the Mississippi NAACP directed its branches to organize black parents to petition local school boards to desegregate the schools. None of these parents could have had the slightest illusion as to the confrontation they were facing, the certainty that it would call forth violence, and the possibility of that violence being visited on their children. Percy Greene, the conservative black editor of the *Jackson Advocate*, called the NAACP's school desegregation crusade "leading a group of unthinking and unrealistic Negroes over the precipice to be drowned and destroyed in the whirlpool of hate and destruction." The Mississippi NAACP begged to differ. The Vicksburg chapter filed the first petition on July 18, calling on the local school board to "take immediate concrete steps leading to the early elimination of segregation in the public schools."[62]

The white power structure responded with the strength of public opinion—and implicit threats. When the *Vicksburg Post* published the names of all those parents who had signed the petition, several asked that their names be removed. By mid-August NAACP chapters in Natchez, Jackson, Yazoo City, and Clarksdale filed similar petitions. As advised by the state conference, the Yazoo City branch asked the school board for "reorganization" of the school system on a "non-discriminatory basis." In each case the Citizens' Council published signers' names, addresses, and phone numbers in large ads in the local newspaper. Reprisals, threats, and intimidation inevitably followed. "If the whites saw your name on the list," said Aaron Henry, "you just caught hell."[63]

Death threats became routine for the petitioners. Many lost their jobs, as did their loved ones. Local banks called in their loans or forced those who signed to withdraw their money. Rocks and bullets flew through their windows. Insurance companies canceled their policies. "In each instance there has been some form of economic reprisal or physical intimidation," Medgar Evers reported to the national office. The number of petitioners who withdrew their signature grew quickly: in Clarksdale 83 of 303; in Vicksburg 135 of 140; in Jackson 13 of 42; and in Natchez 54 of 89.[64]

"In Yazoo City, in particular, signers have been fired from their jobs," Evers reported. "Telephone calls threatening the lives of the persons, have created a continued atmosphere of tenseness which has the city's two racial groups on edge." Fifty-one of the fifty-three signatories of the Yazoo

City petition soon removed their names; the other two had already left town. In fact, even though NAACP membership in Mississippi swelled in 1954 and grew by 50 percent in 1955, the Yazoo City branch, which before the petition had two hundred members, soon ceased to exist.[65] Evers wrote to the national office a few months later, "Honestly, Mr. Wilkins, for Yazoo City there doesn't seem to be much hope. The Negroes will not come together and our former president has not cooperated at all. It appears that [members of the Citizens' Council] have gotten next to him and we can't get any results, not even [to] call a meeting. One thing, the people are afraid—I would say it is worse than being behind the Iron Curtain."[66]

Virtually every black activist in the state had heard the rumor of the Citizens' Council "death list." Arrington High, who put out a small newspaper in Jackson called *Eagle Eye: The Women's Voice*, wrote on August 20, 1955, a week before the murder of Emmett Till, "Citizens' Council members in Leflore County, 'Eagle Eye' alleges, met last Thursday and prepared a list of Negro men to be murdered." High noted that his source "happens to be a peckerwood." According to High, the "Citizens' Council has ordered that no law enforcement body in ignorant Mississippi will protect any Negro who stands upon his constitutional rights." He also directly addressed the white men threatening him: "To white hoodlums who are now parading around the premises of Arrington W. High, editor and publisher of 'The Eagle Eye,' at 1006 Maple St., Jackson, Miss.—this is protected by armed guard."[67] Frederick Sullens, editor of the *Jackson Daily News*, predicted, "If a decision is made to send Negroes to school with white children, there will be bloodshed. The stains of that bloodshed will be on the Supreme Court steps."[68]

It was in this context that a Chicago teenager walked into Bryant's Grocery and had his fateful encounter with Carolyn Bryant. After Till's murder defense attorney J. J. Breland told William Bradford Huie, "There ain't gonna be no integration. There ain't gonna be no nigger voting." He saw the murder as part of a larger struggle: "If any more pressure is put on us, the Tallahatchie River won't hold all the niggers that'll be thrown in it."[69]

11

PEOPLE WE DON'T NEED AROUND HERE ANY MORE

I ndifference to black lives did not begin in the twenty-first century. Nor was Mississippi entirely a foreign country to the rest of America in the 1950s. It simply did not matter to most white Americans what happened to black Mississippians; they did not know and did not want to know, and routine terrorism did not dent that indifference until the Emmett Till case. In the wake of the 1954 *Brown v. Board of Education* decision, for example, one Delta legislator declared that "a few killings would be the best thing for the state." A few judicious murders now, he suggested, "would save a lot of bloodshed later on."[1] In a speech in Greenville the president of the Mississippi Bar Association included "the gun and the torch" among the three main ways to defend segregation.[2] NAACP activists in Mississippi endured scores of acts of intimidation. Whites opposed to school integration and black voting would "put the hit man on you," Aaron Henry recalled of that era. "I could call a roll of the people who died."[3] Reverend George Lee of Belzoni, Mississippi, would be one of the first on any such roll.

In the South's calculation it took only "one drop" of black blood to make a person black. So George Lee, born in 1902 to a white father and a black mother, was black. All Lee knew about his father was that he was a white man who lived in the Delta. Lee's stepfather was abusive, and his mother died when he was a little boy; her sister took the boy in, and he graduated from high school, a rarity at the time.

While still a teenager Lee took off for the port city of New Orleans, two hundred miles south. He worked on the docks unloading banana boats from Guatemala, Honduras, Jamaica, Martinique, the Windward Islands, and the Ivory Coast. In the evenings he took courses in typesetting, hoping to learn a trade that would reward his strong mind rather than employ his strong back. Deep inside, however, even before he left Mississippi, Lee had felt the tug of the Holy Spirit to become a minister. By moving to New Orleans he had "evaded" the call for several years, he later said, but he finally gave in to the Lord. In the 1930s he returned to Mississippi to accept a pulpit in Belzoni, the seat of Humphreys County.[4]

Belzoni was home to four or five thousand people, two-thirds of them black, nearly all of whom lived in stark poverty. Among Mississippi's network of civil rights activists, "Bloody Belzoni" had a reputation as "a real son of a bitch town" where white lawmen policed the color line relentlessly. Whites told a journalist after the *Brown* decision that "the local peckerwoods" in Belzoni "would shoot down every nigger in town before they would let one, mind you, just one enter a damn white school."[5] The white manager of the local Coca-Cola bottling plant told a reporter for the *New Republic*, "If my daughter starts going to school with nigras now, by the time she gets to college she won't think anything of dating one of 'em." Like interracial dating, black voting was also a self-evident abomination: "This town is seventy percent nigra; if the nigra voted, there'd be nigra candidates in office."[6]

Although in 1944 the U.S. Supreme Court had ruled the "white primary" unconstitutional, white Democrats in Mississippi continued to insist that aspiring black voters could be barred from the Democratic primary as though it were a private club. "I don't believe that the Negro should be allowed to vote in Democratic primaries," said Thomas Tubb, the chair of the state Democratic Executive Committee. "The white man founded Mississippi and it ought to remain that way."[7] One banner headline in the *Jackson Daily News* declared, "Candidates Say Delta Negroes Aren't Democrats," which expressed the editors' most presentable public argument against black voting.[8] This was a plain violation of both the Fifteenth Amendment to the U.S. Constitution and a U.S. Supreme Court decision that had been in effect for more than a decade. But the government of the United States failed to see that the national honor was at stake in the white South's open denial of citizens' voting rights. Only the NAACP and small groups of activists objected. Most Americans, North and South, kept silent.

Those who organized Mississippi's effort to block blacks from voting did so without shame. The day before the murder of Emmett Till, Thomas Tubb announced that no African American would ever be allowed to vote in Clay County, where he made his home, "but we intend to handle it in a sensible, orderly manner." Blacks "are better off" not voting, Tubb continued, than being "given a whipping like some of these country boys plan to do."[9] Ten days later Tubb insisted he knew of "no widespread or systematic effort to deny Negroes the voting right," but a day after that denial he appointed a statewide committee "to study ways of cutting down the numbers of Negro voters."[10]

Mississippi had the highest percentage of African Americans in the country and the lowest percentage registered to vote. In the thirteen counties with a population more than 50 percent African American, black people cast a combined total of fourteen votes. In five of those counties, not one African American was a registered voter; three listed one registered African American who never actually voted. In the seven counties with a population more than 60 percent black, African Americans cast a combined total of two votes in 1954.[11] Even so, on April 22, 1954, the Mississippi legislature passed a constitutional amendment explicitly designed to keep black people from the polls: it required citizens wishing to vote to submit a written explanation of the state constitution to the registrar, who would determine whether the interpretation was "reasonable."[12] Seven months later, as the first *Brown* decision rocked the state, Mississippi voters ratified the amendment by a five-to-one margin.[13] The Associated Citizens' Councils of Mississippi, determined to block African American voting at all costs, declared that it was "impossible to estimate the value of this amendment to future peace and domestic tranquility in this state."[14]

To reduce the number of black ballots, the Citizens' Councils had relied mainly on economic pressure, but the message could arrive in considerably starker terms. On July 30, 1955, Caleb Lide, one of the tiny handful of registered voters in Crawford, received an unsigned letter threatening, "Last warning. If you are tired of living, vote and die."[15]

Despite Belzoni's tough side, Reverend Lee's hard work and spiritual depth helped him achieve a good life there. A gifted and fiery preacher, Lee eventually became pastor at three small churches.[16] "Unlike his brethren," wrote the renowned black journalist Simeon Booker, "he preached well beyond the range of Bible and Heaven and the Glory Road."[17] He saw nothing eternal about the Jim Crow social order, and apparently his convictions were infectious. In 1936, when he was thirty-two, he married a steady, quiet twenty-one-year-old named Rose. He and Rose ran a brisk printing business out of the rear of the small grocery store they operated in their house at 230 Hayden Street, in the heart of Belzoni's black community, and Reverend Lee became a leader in the community. "He had a thinking ability better than most of the others," his wife recalled, "so they came to him."[18]

In the early 1950s Dr. T. R. M. Howard recruited Lee as vice president of the Regional Council on Negro Leadership. Lee's eloquent speeches at RCNL rallies became legend. In one much-lauded address to ten thousand black citizens gathered in Mound Bayou he said, "Pray not for your mom and pop. They've gone to heaven. Pray that you can make it through this hell."[19] Booker called Lee "a tan-skinned, stumpy spellbinder" and found his oratory irresistible: "Backslapping the Delta farmers and giving each a sample of his fiery civil rights message, Lee electrified crowds with his down-home dialogue and his sense of political timing."[20] Many of Lee's listeners came to regard him as the most militant preacher in the Delta.[21]

In 1952 and 1953, with the help of Medgar Evers, Lee and his friend Gus Courts, a grocer, organized the Belzoni branch of the NAACP. Courts became the first branch president, and Lee was soon the first black citizen registered to vote in Humphreys County since the end of Reconstruction.[22] They held a number of voter registration meetings, for which Lee printed leaflets. According to Courts and Roy Wilkins, the national NAACP executive director, they managed to register roughly four hundred African Americans. When Sheriff Ike Shelton refused to accept poll tax payments from African Americans and ordered Lee to "get the niggers to take their names off the [registration] book," Lee and Courts threatened to sue him.[23]

This affront brought down the wrath of the Citizens' Council, which launched a campaign of intimidation and reprisal that soon forced virtually all black voters to remove their names from the registration rolls. By May 7 the number of African American voters in the county fell to ninety-two.[24] Local white wholesalers refused to sell Courts wares for his store or to extend him credit. If he didn't take his name off the registration list, they told him, he would lose his lease. No doubt frustrated that economic reprisals were not working, Citizens' Council leaders next assured Lee that if the two men would simply remove their names and cease their registration efforts the Council would protect him and Courts from harm.[25]

The implicit threat was something to weigh soberly. White men had recently beaten two black ministers who had advocated the ballot for African Americans at Starkville and Tupelo.[26] In Belzoni the NAACP's adversaries responded to the successful voter registration drive by smashing the windshields of eighteen parked cars on a single street in the black community and shattering the windows of a number of black-owned businesses. The vandals left a note that promised, "You niggers paying poll tax, this is just a token of what will happen to you."[27] On an evening in the spring of 1955 a mob of white men swarmed Elks' Rest, a local African American social hall, smashed up the place, destroyed equipment, tore up the checkbook, and left this note: "You niggers think you will vote but it will never happen. This is to show you what will happen if you try."[28]

By May 7, 1955, Lee had received countless death threats by phone and mail, including from one anonymous caller who said, "Nigger, you're number one on a list of people we don't need around here any more." That night Lee drove downtown to pick up a suit from the dry cleaner's for church the next day. It was after eleven, but relationships were informal among the town's black merchants and Lee knew that his friend, who lived in the same building where he kept his dry cleaning shop, would not mind pulling a clean suit out of the back room.[29]

Heading home, Lee drove past Peck Ray, a local handyman, and Joe David Watson Sr., a gravel hauler. Both men belonged to the local Citizens' Council. Watson had been arrested recently for shooting into a black sharecropper's home, but District Attorney Stanny Sanders had chosen not to try the case. According to records from the FBI investigation conducted later, witnesses said "they saw two men leave a downtown street corner where they had been standing, enter Ray's green, two-toned Mercury convertible, drive away and return shortly afterward. Several witnesses saw a convertible fitting that description following Lee with only its parking lights on."[30]

As Lee neared his home in a black neighborhood, the convertible pulled up behind him. Witnesses thought one of the passengers in the car looked like Sheriff Ike Shelton.[31] The first shot flattened one of Lee's tires. Then the Mercury pulled up alongside Lee's car, and a .20-gauge shotgun blast blew away his lower left jaw. Lee's car careened into a nearby frame house, collapsing the porch and knocking a huge hole in the front wall. Blood pouring from his head, Lee staggered from the wreckage. A passing black taxi driver saw him collapse and whisked him off to Humphreys

County Memorial Hospital, but Lee died in the backseat. A coroner's jury found that he died from blood loss from wounds caused by about two dozen number-three buckshot.[32] Yet the *Jackson Clarion-Ledger*'s headline the following day was "Negro Leader Dies in Odd Accident." The FBI report on the case noted, "Sheriff I. J. Shelton made public statements that the metal fragments in Lee's jaw were most likely fillings from his teeth." This was a little over three months before the murder of Emmett Till.[33]

Less than an hour after the midnight shotgun blasts took Lee's life, operators for Belzoni's telephone system reportedly began telling black customers that all of the town's long-distance lines were in use. So Lee's friends sped north to Mound Bayou to inform Dr. Howard, who called Representative Charles Diggs in Michigan, who called the White House. Others raced to Jackson to tell Medgar Evers and A. H. McCoy, the president of the state conference of NAACP branches. Evers assembled all the known facts for the national press. "It was clearly a political assassination," recalled Roy Wilkins at the national office, "but the local lawmen practically pretended that nothing had happened."[34] In his annual report for the Mississippi state office Evers was blunt: "[Lee's] independent business, print shop and grocery, made it difficult to squeeze him economically, so their only alternative was to kill him."[35]

McCoy called Ruby Hurley, the NAACP southeastern district director, who was on assignment in Panama City, Florida. She caught the first morning plane to Jackson, where she met Evers, who drove her south to Belzoni. Hurley noticed that Evers was unusually defiant and carried a gun. "Medgar was brand-new then and had some ideas we had to change," she said, adding that he was "anything but nonviolent." Seeing an unmarked sheriff's car on their trail, Hurley decided to keep quiet about it: "I was afraid he might stop and ask the man what he was following us for." In Belzoni they found the black community braced with rage and terror: many residents feared what might happen next, while others openly advocated revenge.[36]

Hurley persuaded Wilkins to have the national office match a $500 reward offered by the RCNL for any information leading to the arrest and conviction of the murderers.[37] Word came to Evers and Hurley that a man named Alex Hudson and a young female schoolteacher had wit-nessed Lee's murder from a porch directly across the street. Hudson had fled to the home of relatives in East St. Louis the very next day.[38] The schoolteacher "moved suddenly from her home during the night and has not been heard from since," an NAACP investigator reported.[39]

While the Mississippi and New York offices of the NAACP sought evi-dence, Lee's family made funeral arrangements. They originally planned to hold the services in the sanctuary of Greengrove Baptist Church, a few blocks from the Lee home; however, it became clear that the church would hold only a fraction of the two thousand mourners. So deacons moved the pews onto the church lawn, while the funeral home director had his men place the casket on the back of a large flatbed truck, which they parked against the rear wall of the brick church, where they built a rough altar and a podium for the speakers.[40]

Rage as well as sorrow gripped the huge crowd as Reverend W. M. Walton opened the proceedings.[41] The service was interrupted several times by shouts of "He was murdered!" The mood was angry, even vin-dictive; Richard West, active in the RCNL and a staunch member of the Greenwood NAACP, attended the funeral carrying a .38 revolver; his wife packed a .32, and his mother carried a switchblade.[42] T. R. M. Howard addressed the crowd, declaring, "We are not afraid. We are not fearful. . . . Some of us here may join him, but we will join him as courageous warriors and not as cringing cowards." Rose Lee had ordered her husband's casket to be open in order to refute the sheriff's claim that he had died in an "au-tomobile accident." Foreshadowing the photographs that would define the Till case, pictures in *Jet* and various black newspapers showed Lee's left jaw all but blasted away and the hundreds and hundreds of mourners.[43] After the funeral NAACP members went back to Belzoni to hold voter registration classes.[44]

On May 22 the NAACP sponsored a memorial rally that packed the Elks' Rest in Belzoni with a crowd estimated at more than a thousand. Wilkins came from New York to speak for the national organization, and

Howard told the audience, "There are still some Negroes left in Mississippi who would sell their grandmamas for half a dollar, but Rev. Lee was not one of them." McCoy challenged the local lawmen who had watched them file into the hall: "Sheriff Shelton is sitting outside that door right now, he and his boys. Come back from his fishing just so he could watch this meeting. I say he might better be investigating the murder of the Rev. Lee than watching this meeting or taking his little tin bucket with some bait in it and going fishing. The sheriff says that the Rev. Lee's death is one of the most puzzling cases he's ever come across. The only puzzling thing about it is why the sheriff doesn't arrest the men who did it."[45]

Except, of course, that it wasn't the least bit puzzling.

Had District Attorney Sanders prosecuted Joe David Watson Sr. for shooting into the home of a black sharecropper, George Lee might not have been killed. By failing to act, Sanders practically green-lighted the murder, which he also declined to prosecute. The message to the community was clear: anyone pushing for black voting rights could be killed with impunity. According to a 1956 FBI memo, Sanders acknowledged that the bureau's investigation of the case "conclusively demonstrates that criminal action was responsible for Lee's death," but he maintained that the identity of the assailants was not "sufficiently established by usable evidence to warrant presentation to the grand jury." To proceed was pointless, he told FBI agents, because a Humphreys County grand jury "probably would not bring an indictment, even if given positive evidence." Besides, race relations in Belzoni had settled down after the murder, and to reopen the case would only stir things up. The Justice Department declined to file civil rights charges, claiming that there was insufficient evidence that the murder bore any relation to voting rights. FBI agents returned to Watson his .20-gauge and shells.[46]

The morning after Lee's murder was Sunday, Mother's Day, but Gus Courts's grocery store was open. After losing his closest friend, it must have felt like a kick in the stomach to see Percy Ford of the Belzoni Citizens' Council leaning over the counter.

"They got your partner last night," Ford reportedly said.

"Yes, you did," Courts replied accusingly.

"If you don't go down and get your name off the register, you are going to be next. There is nothing to be done about Lee because you can't prove who did it." Ford then spoke even more bluntly: "Courts, they are going to get rid of you. I don't know how and I don't want to know how."

The sixty-five-year-old grocer told Ford that he would as soon die a free man as live a coward. In any case, Courts said, he did not intend to take his name off the books. It had taken too much effort to get it there in the first place.

The Citizens' Council was as good as its word. Courts's landlord soon raised his rent so high he was forced to move his store. Planters refused to hire any of the day laborers he hauled in at harvest time with his bus and truck service. Two wholesale grocers canceled his credit. Courts paid cash for two weeks, after which the wholesalers refused to sell him any more merchandise. He had to drive the hour and a half to Jackson himself to buy provisions.[47]

In Humphreys County the Council gave lists of registered African Americans to local white businesses. If an employer or prospective employer found a black person's name on the list, Evers explained, "he'd say, 'we can't employ you until you get your name off this list.'" With this method they knocked registration down to about ninety names, and against this harder core of black voters the Council brought other types of pressure. Evers observed that Sunflower County had lost all 114 black voters due to these methods, and Montgomery County lost all 26. Asked if those numbers represented the goals of the Citizens' Council, Robert Patterson claimed, "We aren't against anyone voting who is qualified under our new registration law."[48]

By the first day of August, the day before the 1955 Democratic primary, only twenty-two African American voters remained on the books in Humphreys County, most of them terrified. "I was notified that very next morning that the first Negro who put his foot on the courthouse lawn would be killed," Courts recounted. Nearly all of the black voters still on the books met in Courts's store later that day to talk over what they should do. "I told them that we would go down to vote, if they were willing to go. They said they were, and we went down to vote."

Courts continued, "After we had gotten to the registrar's office, we were handed a sheet of paper which contained 10 questions. They told us we could not have the ballots until we had answered the questions." The first question asked if they were members of the Democratic Party. "The next question was, 'Do you want your children to go to school with the white children?' The next question was, 'Are you a member or do you support the NAACP?' " The registrar refused to accept any of their votes. Not one black ballot was counted in Humphreys County in 1955.[49]

Subsequently Courts and four of his comrades wrote and sent a petition to Governor Hugh White asserting that their voting rights were being denied and their lives threatened and asking for protection of both their ballots and their bodies. They sent a copy to the U.S. Department of Justice. The response was chilling. "Now, the Governor sent that petition back to Belzoni to the White Citizens' Council," Courts explained. Percy Ford showed the original petition to all five signatories, one by one. "You signed this petition and sent it to the governor," Courts remembered Ford telling him. "Now you see how much protection you have gotten from the Governor."[50]

As the coercion increased, Courts scrawled a note to Evers: "I am reporting a few the Citizen Council putting the preasher on. Mr. V. G. Hargrove, Tchula, Misss. Farmer. Mr. Neely Jackson Tcula Miss. farmer. Mr. Fread Myls Belzoni Miss. merchen, Mr. Will None, planter, Belzoni, Miss., Mr. Willie A Harris, Taxie Cab oner Belzoni Miss. Mr. Gus Couters Belzoni merchen and a lot of others I dount have thire names. All these men have ben ask to get out of the NAACP. And to return thire poll tax or tare up receat. Your truley Gus Courts, 61 First St. Belzoni Miss."[51] Evers filed the letter along with several similar reports. Citizens' Council members persisted in warning Courts to get out of Belzoni or face the consequences.

About eight o'clock on Friday night, November 25, 1955, a forty-two-year-old black woman named Savannah Luton entered Courts's grocery store on First Street in Belzoni. "I was busy waiting on customers in my store," Courts recalled. He greeted Luton from behind the old drink box and cash register. "I was buying a dime's worth of coal oil and I heard this noise that sounded like firecrackers," she said later. "I bent down to look out the window and told Mr. Courts, 'There's some white folks out here shooting at us.' He didn't even know he was shot until then."

The shotgun had been loaded with unconventionally large buckshot, and two quick blasts splattered the upper edge of the bed of Courts's pickup truck and sprayed irregular holes across the store's plate glass window. When Courts touched his side his hand came away covered with blood; the slugs had caught him in his left arm and stomach. Unharmed, Luton rushed out the front door just in time to see a two-toned automobile spraying dust and gravel as it raced away toward downtown Belzoni. There was a white man at the wheel and other people in the car whom Luton couldn't make out. Running back into the store, she found Courts fallen. He had one hand pressed up against the wound to his side and the other to the wound in his arm. Blood seeped through his fingers and dripped onto the wood floor.[52]

Friends and family rushed Courts to the hospital—not the local hospital nearby but the Taborian Hospital in Mound Bayou. "When I walked out to get into the car, I told my friend, Ernest White, who came to take me to the hospital, that I wanted to go to Mound Bayou hospital, which is 80 miles away. The sheriff came over in 30 minutes to my store, after I had left for the hospital. He asked my wife where I was. He said he had been over to the hospital, just two blocks away, for 30 minutes waiting for me." When his wife said he'd gone to Mound Bayou, the officer was not pleased. "I believe they would have finished me off if I had landed up in the Belzoni hospital," said Courts.[53] Sheriff Shelton complained: "They took Courts across two counties, to 80 miles north of here, though we have the best hospital in the world and two of the best doctors."

"I've known for a long time it was coming and I tried to get prepared in my mind for it," Courts said, "but that's a hard thing to do. It's bad when you know you might get shot just walking around in your store." Although it was more than a year before Courts had full use of his arm again, he recovered from the buckshot. Other wounds proved harder to heal. No one was ever charged in the case. The *New York Times* ran one tiny story about the murder of Reverend Lee, but no national press reported on what happened to Courts or on the wave of intimidation against black voters all over the state. U.S. Attorney General Herbert Brownell Jr. insisted that under federal law the Justice Department had no authority to prosecute, even though the courts had long since found that the Fourteenth Amendment to the U.S. Constitution created a national citizenship and so empowered the federal government to protect the right of all citizens to vote.[54]

Like Lee and Courts, a sixty-three-year-old African American cotton farmer named Lamar Smith decided that he would risk everything to help bring the vote to black Mississippians. About two weeks before Milam and Bryant drove to Reverend Wright's farmhouse to take Emmett Till, Smith went to the courthouse in Brookhaven, Mississippi, to obtain more of the absentee ballots he was distributing to African Americans so that they could vote without being intimidated or attacked.[55] It was ten o'clock on Saturday morning and the square was filled with people. At least three white men set upon the unarmed Smith as he crossed the courthouse lawn and beat him mercilessly. Then at least two of them held him while another fired a .38 revolver into his heart and, by one account, fired a second shot into his mouth. Dozens of people stood nearby. The sheriff was close enough to recognize at least one of the killers and to describe the bloodstains on the shirt of another. The FBI investigation stated flatly that his assailants killed Smith "in front of the sheriff."[56]

Arrington High's *Eagle Eye* newspaper claimed that the dozens of witnesses to the murder were "ordered to shut up their mouths." He angrily demanded, "Was this man murdered by elected officials?"[57] The front page of the *Jackson Advocate*, a conservative black newspaper, declared that Smith's murder was "generally regarded as resulting from sentiment created against Negro leaders in the state by the White Citizens' Councils."[58] Even Brookhaven's own Judge Tom Brady acknowledged that there had been no trial in the Smith killing because no white man was willing to testify against another in the murder of a black man.[59]

Not all whites remained silent, however. When the sheriff would not make an arrest, despite personally witnessing the murder, District Attorney E. C. Barlow tried unsuccessfully to persuade the governor to send highway patrol officers to investigate, calling the murder "politically inspired." The chairman of the all-white local grand jury complained bitterly in a lengthy statement to the newspapers, "Most assuredly somebody has done a good job of trying to cover up the evidence in this case, and trying to prevent the parties guilty, therefore, from being brought to justice." Even though "there were quite a number of people alleged to have been standing around and near said killing, yet this Grand Jury has been unable to get the evidence, although it was generally known or alleged to be known who the parties were in the shooting." Claiming to speak on behalf of the whole grand jury, the chairman raged, "We think it impossible for people to be within 20 or 30 feet of a difficulty in which one party is shot, lost his life in broad, open daylight, and nobody knows nothing about it or knows who did it."[60]

Despite the openly political nature of the Mississippi attacks, the national news media soft-pedaled the murders of George Lee and Lamar Smith and said little about the attempted murder of Gus Courts. So it's easy to understand why the murderers and those who sympathized with them would think that the country didn't care about the rights or even the lives of African Americans in Mississippi. And it is no surprise that J. W. Milam and Roy Bryant would assume they could murder Emmett Till without real consequences.

After Courts was released from the hospital in Mound Bayou, his wife and Evers persuaded him to relocate his family to Chicago, where the NAACP set him up in a small grocery business on the South Side.[61] Gunmen sprayed the Jackson home of A. H. McCoy with bullets. Dr. C. C. Battle, who had performed the autopsy on George Lee, fled for Kansas City. Dr. Howard sold his home and nearly eight hundred acres and moved, first to California and then to Chicago, where he opened a medical practice and went into politics. They all felt sure they would be killed if they stayed in Mississippi.[62] Amzie Moore wrote to a Chicago friend in late 1955, "It's ruff down here and a man's life isn't worth a penny with a hole in it. I shall try to stay here as long as I can but I might have to run away up there. Look for me right after the first of the year."[63] Moore managed to stay, but his house became a virtual arsenal and was lit up like Christmas every night of the year.[64]

Courts never returned to Belzoni. "You see before you an American refugee from Mississippi terror," he testified two years later before the U.S. Senate Committee on the Judiciary. "We had to flee in the night. We are the American refugees from the terror in the South, all because we want to vote."[65]

Rumor had it that friends had smuggled him out of town in a casket.[66]

12

FIXED OPINIONS

On the Sunday afternoon before the trial of Roy Bryant and J. W. Milam began, C. Sidney Carlton, one of the five attorneys defending them, visited the chief witness for the prosecution. Carlton, a round-faced, bespectacled man well into middle age, would soon become the president of the Mississippi Bar Association.[1] The kindest term for what he was about to do is *witness tampering*. He knocked at the door of the unpainted tenant farmhouse where Reverend Wright lived, the same door that Milam and Bryant had pounded on when they came to snatch the black boy from Chicago. Carlton had come to warn Wright about testifying against his clients.[2]

Like much about the trial of Emmett Till's murderers, there was a performance quality to Carlton's visit. His warning was superfluous. After meeting the heavily armed minister at his home, Moses Newsome of the *Memphis Tri-State Defender* wrote, "[Moses Wright] seems certain he can handle any situation that comes up while he is awake. Each time that cars slowed down in front of the house, he kept telling this reporter, 'Don't worry. It is all right here.' "[3] Wright avoided sleeping at his house, however, as he waited for his chance to testify in court; instead he slept in his car in a rural cemetery or another secret location. "I spend some nights here and some nights I don't," he told reporters. "I'm superstitious." Every day, as he and his sons picked his twenty-five–acre cotton crop, his shotgun was nearby. He also kept a rifle close at hand. The boys were living with relatives.[4]

Wright knew to a certainty who had abducted his nephew, tortured and butchered him. He knew how remarkable it was that there would even be a trial and knew to a near certainty that the murderers would be found not guilty by a jury of white men. He knew that African Americans had been killed by whites for centuries without real consequence. There was no reason to imagine that this time would be any different. But like George Lee, like Gus Courts, like countless other African Americans pasting bumper stickers to their cars, attending rallies, signing their names on school integration petitions, and attempting to register and vote, he chose courage. That was why he stayed in the Delta. "There was not a trace of fear about [Wright] as he asked his visitors to enter," a reporter visiting the reverend's home wrote later. "There was even a little defiance when Carlton suggested that things might not go well with Moses Wright if he fingered Milam and Bryant."[5]

Wright had to wait only three weeks. The trial began on Monday, September 19, a mere twenty days after the sheriff's deputies fished Emmett's bloated body from the Tallahatchie River. This left little time for a proper investigation, which was the point. Sheriff Strider of Tallahatchie County, who had successfully claimed jurisdiction and so was responsible for that investigation, had in fact urged the judge to begin the trial only a week after they found the body even though he had not found any evidence or witnesses.[6] "We haven't been able to find a weapon or anything," he told reporters. Nor did the state have any credible notion of where the murder had taken place.[7] This lack of information would cast a lingering shadow

over the question of Strider's jurisdiction, since the kidnapping clearly occurred in Leflore County.

In fact, it was Leflore's Sheriff George Smith who had arrested Milam
and Bryant. Had he been given jurisdiction, the effect on the trial might
have been different. No less a pillar of Mississippi civil rights politics
than Dr. T. R. M. Howard called Sheriff Smith "the most courageous
and the fairest sheriff in the entire state of Mississippi." Ruby Hurley of
the NAACP's southeastern regional office also sang Smith's praises.[8] But
Strider made his claim based on the discovery of the body about ten miles
into his county. He also claimed that he had found some blood on a bridge
that indicated where the body had been dumped. The FBI lab soon determined that it was not human blood, but that did not matter; Strider got
jurisdiction anyway.

And Strider was determined to maintain control of the proceedings.
Another reason for the hasty trial was that Strider's four-year term ended
in three months' time; postponement until the next session of court would
allow Strider's successor to take over the leading law enforcement role.[9]

Henry Clarence "H.C." Strider was a tobacco-chewing, cigar-smoking
former football player with a bad heart and a gruff demeanor.[10] The
owner of 1,500 acres of prime cotton land, Strider farmed his plantation
with 35 black sharecropping families and kept a general store and filling
station on the property. He also operated a crop-dusting company with
three airplanes.[11] Seven tenant houses lined the driveway to his home and
each had one huge letter on its roof, visible from the highway or one of
his planes: S-T-R-I-D-E-R.[12] "He was like a godfather over the Delta,"
Carolyn Bryant recalled. "All of the other sheriffs and police departments
and all, whatever Strider said, that's what they did."[13] Elected in 1951,
he was "as tough a sheriff that's ever been around here," recalled Crosby
Smith, Emmett Till's uncle. "He weighed about three hundred pounds and
he walked heavy."[14] If he carried a gun in the courthouse during the trial it
was not visible, but an oversized blackjack protruded prominently out of
the right front pocket of his trousers.[15]

Sheriff Strider ruled at the Tallahatchie County Courthouse, a badly aging three-story brick castle erected in 1915 in Sumner.[16] Built around the courthouse square half a mile from the highway, Sumner, one of two county seats in Tallahatchie, had a population of nearly six hundred, more than two-thirds of them black and not one registered to vote. Strider kept security for the trial tight. "Deputies wearing gun belts ambled in and out, as if it were the set of a TV western, and frisked everyone who entered the courtroom," wrote Dan Wakefield for the *Nation*.[17] "I have received over 150 threatening letters and I don't intend to be shot," Strider announced to the press. "If there is any shooting, we would rather be doing it."[18]

To judge from the initial press coverage, much of white Mississippi shared Chicago's outrage at Till's murder. On September 1, the day after his body was found, the *Clarksdale Press Register* called it "a savage and useless crime" and stated flatly, "If conviction with the maximum penalty of the law cannot be secured in this heinous crime, then Mississippi may as well burn all its law books and close its courts." The *Greenwood Commonwealth* ran a front-page editorial asserting, "Citizens of this area are determined that the guilty parties shall be punished to the full extent of the law." The *Vicksburg Post* called it a "ghastly and unprovoked murder" and urged "swift and determined prosecution." Governor White dispatched telegrams to District Attorney Gerald Chatham urging energetic prosecution of the case and to the national office of the NAACP promising that the Mississippi courts "will do their duty."[19]

T. R. M. Howard, the civil rights leader and surgeon from Mound Bayou, was on a business trip to Chicago when he heard that Till's body had turned up in the Tallahatchie. "There will be hell to pay in Mississippi," he told reporters.[20] Mamie Bradley also fired on the Magnolia State, vowing, "Mississippi is going to pay for this."[21] Much to the chagrin of most white Mississippians, Mamie called what had happened to her son "an everyday occurrence" there, and said visiting Mississippi was "like walking into a den of snakes."[22] In New York, Roy Wilkins made an even more vehement condemnation, which appeared in newspapers across the nation and on the front page of the *Jackson Clarion-Ledger*. "It would appear by this lynching that the State of Mississippi has decided to maintain white supremacy by murdering children," he said. "The killers of the boy felt free to lynch him because there is in the entire state no restraining influence of decency, not in the state capital, among the daily newspapers, the clergy nor any segment of the so-called better citizens."[23]

These statements would linger in the minds of white Mississippians. Their outrage at the Till murder dissipated as they began to smart at all the criticism directed at their state, and by implication at them. Within days a full-scale backlash began to roll. White Mississippians resented the sweeping condemnations of their state in the Northern press and particularly the excoriating denunciations by Wilkins and other civil rights leaders. Many Mississippi editors began to fire back indiscriminately.[24] The editor of the *Picayune Item* snarled that a "prejudiced communistic inspired NAACP" could not "blacken the name of the great sovereign state of Mississippi, regardless of their claims of Negro Haters, lynching, or whatever."[25]

Some used the searing national criticism of Mississippi to explain the performance of justice unfolding in the courtroom: the state's critics, they argued, had swelled Strider's sympathy for Roy and Milam. But prosecutor Hamilton Caldwell countered that Strider "was for the boys all along." Carolyn Bryant later noted that well before the murder the Milams and Bryants believed that their association with Strider made them immune to prosecution. As early as September 3, only three days after Till's body was found, Strider told reporters he did not believe the body was Till's: "The body we took from the river looked more like that of a grown man instead of a young boy. It was also more decomposed than it should have been after that short a stay in the water."[26] This contrasted sharply from his assessment of the body's condition right after they pulled it from the river, when he said it looked like it had been there only two days. This shift had a political purpose, of course; if it had been in the water for more than a few days, the body could not have been Emmett Till's. The following day Strider told reporters, "The whole thing looks like a deal made up by the NAACP."[27] Sheriff George Smith of Leflore County quickly and publicly disagreed. His deputy, John Ed Cothran, could not keep silent, either, and stated emphatically that he'd been present when Moses Wright identified the body by the silver ring engraved with Till's father's initials.[28]

But in the Tallahatchie County Courthouse their dissents amounted to little. Strider soon elaborated on his theory to reporters from the *Greenwood Morning Star*: "It just seems to me that the evidence is getting slimmer and slimmer. I'm chasing down some evidence now that the killing might have been planned and plotted by the NAACP."[29] Hodding Carter Jr., the editor of the *Delta Democrat-Times*, wondered, "Whoever heard of a sheriff offering on the flimsiest construction of fact, the perfect piece of evidence for the defense? Without a corpus delicti, there can be no murder conviction—of anyone." Carter pointed out the irony that the same man who now denied that the body was Till's had claimed jurisdiction over the case on the basis of some blood he said he had found on a bridge.[30]

Of course Strider wasn't brazen in a vacuum. The tide of public opinion seemed to run ever more in his favor. Governor White wrote to a colleague, "I'm afraid the public has become so aroused over the NAACP agitation that it will be impossible to convict these men."[31] Though editors and politicians like White blamed the turn on the NAACP and Northern critics, the fuming of Mississippi editors pushed public opinion considerably. Nevertheless, to the genuine surprise of many observers, a grand jury issued indictments against Milam and Bryant for murder on September 7.[32]

The man running the murder trial inside Strider's citadel would draw high marks from the press corps. James Hicks of the National Negro Press Association and the Afro-American News Service, who came to the trial a self-described "skeptic of Mississippi's white man's justice," wrote afterward that no judge "could have been more painstakingly and eminently fair in the conduct of the trial than Judge Curtis M. Swango of Sardis, Mississippi."[33] Greenville's *Delta Democrat-Times* averred, "[Swango] is providing the South with the best public relations it has had since the invention of the Southern Accent first enchanted Northern ears."[34] Murray Kempton of the *New York Post*, immune to any charming drawl of Southern-fried public relations, called Judge Swango "a quiet man firmly and graciously committed to a fair trial whatever the verdict."[35] Even the rabidly segregationist *Jackson Daily News* agreed that the jurist was "to be warmly commended for his scrupulously correct conduct in the face of what must be a difficult situation." That paper also called his selection "good casting": "[He] looks like Hollywood's idea of what a judge should look like."[36]

Though he banned broadcasts, photography, and recording during testimony, Judge Swango delighted journalists by permitting them to shoot pictures during the fifteen minutes before trial and during intermissions. He also permitted smoking in the courtroom and made a gesture toward comfort by suggesting that the men remove their jackets in the sweltering heat.[37] He sipped a cold Coca-Cola during jury selection and let participants and spectators do likewise. Others drank beer without reprimand.[38] On the first day of the trial one of the courthouse custodians walked in with a wooden crate full of glass bottles of ice-cold Coca-Cola and quickly sold each one for a dollar, though at that time soft drinks cost a nickel. When the crate was empty he sold that, too, for a dollar, as a makeshift chair.[39] Yet nobody described Judge Swango's courtroom as lax. "The dignified magistrate has an authoritative voice and a husky gavel and doesn't hesitate to use it," wrote Harry Marsh, a Southern liberal with the *Delta Democrat-Times*. "But most of all he has displayed great patience and fairness in the two days required to select a jury in the hot, crowded courtroom."[40]

The photographers and reporters, especially the African American contingent, appreciated Swango's letting them roam the courtroom before the proceedings and during breaks. One of them, Ernest Withers, snapping pictures for the *Memphis Tri-State Defender*, would become a famed civil rights photographer and seemed at ease even in the tense atmosphere. One day a white man in the crowd suddenly jumped up and demanded, "Nigger, don't take my picture." Withers, who was born in Memphis and had worked as one of the city's first black police officers, shrugged. "Don't worry," he deadpanned, "I'm only taking important people today." James

Hicks, a Northerner, taking nervous drags on a cigarette at the card table set aside for African American reporters, muttered to Withers, "Man, you'll get us lynched down here."[41]

The spectators in shirtsleeves, the sweating bottles of Coca-Cola, the roaming interracial pool of reporters, the judge's demeanor—none of these influenced the substance of the murder trial of Roy Bryant and J. W. Milam. On Monday, September 19, the first day of the trial, selection of ten of the twelve needed jurors began promptly at nine o'clock. Sheriff Strider and Sheriff-elect Harry Dogan, who between them knew virtually everyone in the county, helped the defense lawyers vet the 125 potential jurors. That it would be an all-white jury went without saying. "No Negroes will serve on the jury," explained the *Jackson Daily News*. "Women do not serve on Mississippi juries, either."[42]

The prosecution dismissed three juror prospects for admitting to having contributed money to the defense fund, which reportedly collected $6,000 in the Delta.[43] Other common causes for dismissal included being related to the defendants or one of the attorneys in the case, knowing the defendants, living near where the murder presumably occurred, or holding "fixed opinions" on the case.[44] The prosecution eliminated only one prospective juror, a cotton farmer, for reasons associated with racial prejudice; he did not admit to such prejudice but appeared unable to understand the question or reluctant to respond.[45] Another man, asked whether he had a "fixed opinion" about the crime, replied, "Anybody in his right mind would have a fixed opinion."[46]

On Tuesday morning, the crowded courtroom watched attentively as the defense and prosecution set out to select the two more jurors necessary to proceed with the trial. They did not complete the process until almost eleven o'clock; the court had to call nine more prospective jurors in order to confirm the last two. The prosecution rejected six for having contributed to the defense fund of Milam and Bryant and used one of its peremptory challenges to block another. In the end nine cotton farmers, two carpenters, and an insurance salesman were selected: all men, all white.[47] The *Greenwood Morning Star* described the local men as "a jury mostly composed of open-collared, sunburned farmers," although the *Memphis Commercial Appeal* noted that one juror wore a necktie that first day.[48] Judge Swango selected J. A. Shaw, one of the nine farmers, to chair the jury.

The court sequestered Shaw and his compatriots at the Delta Inn, a hotel about a hundred yards from the Sumner courthouse. They took their meals in the hotel dining room and received $5 a day besides. Barred from watching television, reading newspapers, or listening to the radio, the members of the jury were also barred from discussing the case with anyone. This did not dissuade local members of the Citizens' Council from calling on jurors individually to ensure they voted "the right way."[49]

After jury selection was complete, Murray Kempton of the *New York Post* observed, "the defense was exuding its satisfaction and its assurance of a two-day trial and a two-minute acquittal."[50] Any jury drawn from Tallahatchie County would have presented a challenge for the prosecution, but this one defied the prospect of a successful outcome. Ten of the twelve hailed from the Mississippi hill country, where race relations were especially harsh. The defense attorneys, in no small measure assisted by Sheriff Strider and Sheriff-elect Dogan, knew enough of the jurors personally to be confident of acquittal. "After the jury had been selected," said senior defense counsel Breland, "any first-year law student could have won the case."[51]

The one fixed opinion that everybody from Tallahatchie County seemed to share was that the jury would find the accused not guilty.[52] This did not detract from the high suspense and absorbing interest of the courtroom drama. Perhaps as many as four hundred people packed the place, and most watched the proceedings with rapt attention, though two deputies played checkers in the jury room throughout much of the trial.[53] The white spectators were mostly farmers. About forty African American observers occupied half a row and often some of the wall space in the rear. Only about fifteen or so of the spectators, black or white, were women.

Merely feeding and housing all the people attending the trial placed a heavy strain on the meager local facilities. Sumner had no real restaurant except a small hotel dining room on the courthouse square, where the jury took their meals. By Monday afternoon a café owner from Clarksdale, twenty-one miles away, had set up a concession stand in the courthouse lobby and passed the word that he would be selling half-chicken box lunches on Tuesday. Most of the white journalists ate at a drugstore across the street that had never sold food before but stocked supplies of sandwiches and soft drinks during the trial. Coca-Cola prices doubled, even when all the cold ones were gone and people sipped from tepid bottles until the refrigerators could catch up. Three blocks away the jukebox blared from Griffin's, where the owner and her employees hawked food and drinks to African American spectators and journalists.[54] The conspicuous interracial group from Louisiana took their lunch on the courthouse lawn.

Even more conspicuous was the sheer size of the press corps. Sumner, a sleepy village of six hundred, was playing host to nearly a hundred journalists and thirty photographers, most of them from distant states and a few from other countries—New York, Chicago, Memphis, Detroit, Miami, Atlanta, New Orleans, Pittsburgh, Toledo, Washington, Ontario, London— and from all over Mississippi: Jackson, Clarksdale, Greenville, and Greenwood.[55] *Time*, *Newsweek*, *Life*, the *Nation*, *Jet*, *Ebony*, and several other magazines sent reporters. Newspapers in Jakarta, Copenhagen, Düsseldorf, Paris, Istanbul, Rome, and Stockholm, among others, showed keen interest. Bill Stewart, who broadcast a radio program several times a day to stations in Louisiana, Minnesota, and Ohio, said, "This is the biggest thing we've ever done; we've got more phone calls from our listeners thanking us for having a man on the scene than anything we've done."[56] On the courthouse lawn, writes the historian Robert Caro, there sprung up, "if not a forest, at least a small grove of tripods supporting television cameras." Three major television networks chartered airplanes that set down in a field seven miles away to pick up film every day and whisk it back to New York.[57] National print journalists wired in their stories; Western Union set up a special booth in Sumner that sent nineteen thousand words on Monday, twenty-two thousand on Tuesday, and far more on Wednesday, Thursday, and Friday.[58] Each time court recessed, radio journalists from Memphis, New York, Detroit, Chicago, Hattiesburg, Jackson, and other cities rushed to the few telephone booths to record their stories for immediate broadcast.[59]

For reporters from London, New York, Chicago, Washington, and even more distant places, Sumner, Mississippi, must have seemed a foreign country.[60] "You lie in bed at night listening to the hounds baying," Dan Wakefield reported, "and during the day you see more men wearing guns than you ordinarily do outside your television screen. I am not ashamed to confess that I was afraid."[61]

Outside a thousand people besieged the courthouse square.[62] African Americans sat on the toasted grass beneath a Confederate statue dedicated to "the Cause That Never Failed," and whites gathered on and around benches across the lawn.[63] Observers noted a violent tension in the crowd. "It was like watching a community you thought you knew reveal itself as something else," said Billy Pearson, a young white man who had been away at the University of North Carolina and come home to run the family farm. Pearson claimed to be appalled by the threatening atmosphere. Sheriff Strider had hired a number of youthful special deputies whom Pearson called "bully-boys" with long sideburns and big pistols who enjoyed pushing people around, he said, especially the black people.[64]

The deputies could not intimidate everyone outside the courthouse. Frank Brown, a labor organizer from Chicago, spent a day there mingling with other black men. Many carried guns, he recalled, and none seemed cowed by the deputies. "Used to be they would charge us with clubs and chase us off the grass," one of the men told Brown, "but they know we ain't running no damn where this time."[65] On Monday afternoon a young black man openly brandished an automatic on the courthouse lawn but jumped into his car and drove away before the deputies could question him and find out who he was.[66] Amzie Moore recounted, "The tension was so thick until, as the blacks and whites mixed on the courthouse grounds, you just looked for an explosion just any time."[67]

One possible spark was an unlikely group of visitors from the United

Packinghouse Workers Association, a union with an increasingly strong commitment to civil rights for all Americans; the UPWA would become a vital part of the national civil rights coalition that emerged in the wake of the Till lynching. The union had sent an interracial delegation from Gramercy, Louisiana, a small town with a union sugar refinery. It was in the grips of a strike and had recently been the site of a women's conference, which had adopted a resolution denouncing the Till murder and calling for justice. "We are building a new, free, unafraid South," the resolution declared.[68] The UPWA sent two white field representatives, two white program coordinators, the African American president of the Women's Auxiliary, and three white wives of striking workers. Sugar workers from rural Louisiana certainly knew how bizarre it was for them to travel in a mixed group. "The fact that Mrs. Lillian Pittman, a Negro, was in our company caused us to have innumerable problems on our trip to Sumner by auto," one reported, without going into details. She also mentioned the "shocked glares from the white populace" in Sumner.[69]

"We motored to Mississippi with Mr. Telfor," Pittman reported. "Afterward Marjorie Telfor, Grace Falgoust and Mrs. Vicknair and I sat down to eat lunch under a tree in the shade. The three white ladies and I were sitting down and along came a white photographer and took our picture." He assumed they were from Chicago, but they told him, "'No, we are from Louisiana.' He couldn't believe it because the people in Mississippi don't mix." An older white man nearby suggested that it would be better for the white women to stay away from black people. "I asked the man did he own Mississippi," said Falgoust. "He couldn't answer me."[70]

At some point Pittman spoke to some African Americans hanging around the courthouse about "political action." They were speechless. "Then some answered me, 'Lady, do you want us to be killed and put in the Tallahatchie River?' They also said they were not allowed to vote. To some of them that [the whites] let register, they said, 'Politics are for the whites.' And '[Negroes], on the day to vote, you had better not appear at the polls.'"[71]

For the first day and a half of the trial, the women of the Gramercy

UPWA could not get seats in the courtroom, Pittman reported: "But we tried to make up for it by trying to get as much news as we could out of the courtroom."[72] They interviewed townspeople, handed out leaflets condemning the murder and calling for justice, and "issued press releases to news reporters from all over the country." According to the UPWA delegates, "A great deal of attention has been given to the presence in this tense situation of a friendly, interracial delegation."[73]

In their exchanges with local whites, the women learned much. Pittman overheard two excused jurors acknowledge to one another that they had purposely given answers sure to get them off the jury because they knew "that the defendants had killed the boy, and they did not want to be party to the verdict of 'not guilty,' which they knew would be expected."[74] As they interviewed local people, the women discovered that none doubted the guilt of the accused, but all of them added something like "The jury knows better than to do anything to them."[75]

On their second day at the trial Pittman managed to get into the courtroom, so the three white women left the grounds and walked through Sumner, talking with whoever would talk to them. All the white people they met were hostile to the prosecution. Several, wrote Freida Vicknair, "insisted that the body sent to Chicago was not that of Emmett Till." Contradicting this assertion was the common understanding that the accused were guilty: "No one protested the innocence of Bryant and Milam. In fact, we were told that this crime was justified." The locals were certain of acquittal and believed that in the unlikely event any jurors voted otherwise, they would pay for it with their lives.[76]

Even stranger to the town of Sumner than the interracial UPWA delegation was the group of black reporters patrolling the small town. There was James Hicks of the Afro-American News Service and the National Negro Press Association, who was pivotal not only in covering the trial but in uncovering hidden witnesses, some of whom testified to considerable effect. Simeon Booker and Clotye Murdock were there for *Ebony*, along with their photographer, David Jackson. L. Alex Wilson of the *Memphis Tri-State Defender* accompanied photographer Ernest Withers, who created lasting images of the notorious trial. The *Chicago Defender* probably slung more ink from Sumner than any newspaper, black or white. Reporters William B. Franklin and Steve Duncan and publisher Nannie Mitchell of the *St. Louis Argus* attended. All the other major Midwestern black newspapers—the *Kansas City Star*, the *Cleveland Call and Post*, and the *Michigan Chronicle*—covered the trial.[77] The very sight of white and black reporters greeting one another and exchanging notes in a friendly manner shocked the Sumner crowd. Therein was some of the trial's actual drama, for if almost everyone involved could predict the trial's verdict, few could predict its consequences. The *New York Post* columnist Murray Kempton, for one, thought the locals' reaction revealed "more incredulity than menace."[78]

This wasn't true for the sheriff. Strider blamed all the national hullabaloo on outsiders and seemed to focus on the black press as a prime example. At the outset of each day he would walk, his blackjack protruding from his front pocket, past the table where the black press huddled and offer a cheerful "Good morning, niggers."[79] He interfered with their work whenever he could, reserving them only the barest minimum space, and then only at the direction of Judge Swango. "They allotted us chairs at the Jim Crow press table but during the noon recess while we were trying to get our stories filed in a Negro restaurant"—a pool hall, really—"the crowd would come in and take chairs from our table. I stood up more often than I sat down," complained James Hicks. "We never have any trouble," Strider told television reporters during a break, "until some of our Southern niggers go up North and the NAACP talks to them and they come back here."[80]

Meanwhile Moses Wright slept with a shotgun and harvested his cotton crop, waiting for the chance to speak his truth.

13

MISSISSIPPI UNDERGROUND

U ntil he left Mississippi in late 1955, Dr. Theodore Roosevelt Mason Howard felt sure that his name topped the Citizens' Council "death list" on which the names of his late friends George Lee and Lamar Smith had once appeared. So he kept a .357 Magnum revolver on one bedside table and a .45 semiautomatic on the other and a Thompson submachine gun at the foot of his bed. A rifle or shotgun stood in all four corners of the bedroom and in every other room in the house. His large and well-appointed home and outbuildings would be more properly called a compound; the driveway featured a guardhouse in which armed men sat twenty-four hours a day. He was not a violent man, but he intended to sleep in peace on his own property. That is why, when an African American farm worker named Frank Young showed up at the gate at midnight on the eve of the Till murder trial, the guards were reluctant to wake Howard up. But Young insisted that he had an important story to tell about Till's murder and refused to talk with anyone but Dr. Howard.

In the front room of Howard's house Young told a harrowing story that he had walked and hitched rides for eighty miles to tell: he had witnessed and knew others who had witnessed events in the murder of Emmett Till. Their accounts changed the narrative of the murder significantly, including moving the crime from Tallahatchie to Sunflower County and tying Bryant and Milam directly to the killing.

Early on that Sunday morning of August 28, said Young, three or four black workers had seen a green and white Chevrolet pickup pull onto the plantation in Sunflower County managed by Leslie Milam. Four white men were in the cab of the truck, and in the back Emmett Till sat between two African Americans, Levi "Two Tight" Collins and Henry Lee Loggins, who both worked for J. W. Milam. The pickup stopped in front of a small barn or equipment shed and the group went in. Soon thereafter Young and the others heard the unmistakable sounds of a vicious beating. When Young and another witness snuck up closer to the shed they saw J.W. walk out and get a drink of water from the well. Someone then drove the truck into the shed, and the witnesses watched as it came back out with a tarpaulin thrown over the bed. Emmett Till was no longer visible. All of these witnesses were available to tell their stories, Young told Howard.

Howard had already given over his home as a safe house and headquarters for witnesses, journalists, and crucial visitors like Mamie Bradley and Representative Charles Diggs. He would bring Mamie from Chicago at his own expense and escort her and other witnesses to and from the Sumner courthouse in a well-armed caravan. Diggs had been a guest at Howard's farm on several occasions for the huge annual gatherings of the RCNL, where he was a favorite speaker. But Howard's efforts extended beyond providing a safe haven: he had been leading the "Mississippi underground" that undertook the most effective investigation of the Till murder, helping to find, interview, and eventually protect and relocate several key witnesses.[1]

On a tip, one member of the underground, the reporter James Hicks, had gone to a joint called King's in Glendora, the little crossroads where

J. W. Milam lived. "The place was filthy and the cotton pickers who were enjoying their Sunday off crowded it to the doors," he wrote. Hicks drank beer, danced, and eventually unearthed rumors that Sheriff Strider had locked away two men, Levi "Too Tight" Collins and Henry Lee Loggins, in the Charleston jail to keep them from testifying. Collins and Loggins might well be the two black men Young had seen with Till in the back of the pickup.[2] The rumors would eventually be proven true; defense attorney Breland later confirmed that Strider kept the two men in the jail under false identities both before and for the duration of the trial.[3]

Also in the Mississippi underground were Ruby Hurley, Medgar Evers, and Amzie Moore, who had been investigating the Till case for some time. Myrlie Evers wrote later, "Medgar and Amzie Moore, an NAACP leader from Cleveland, Mississippi, set off from our house one morning with Ruby Hurley, down from Birmingham to investigate. . . . All of them were dressed in overalls and beat-up shoes, with Mrs. Hurley wearing a red bandana over her head. Moore had borrowed a jalopy with license plates from a Delta county. Dressed as day laborers, they made their way among sharecroppers' cabins and cotton fields, looking for people who might know something about the murder."[4]

After his midnight conversation with Young, Howard called some of his colleagues to share the new evidence. Many of the black reporters were already at the house, including Hicks, Simeon Booker, and Robert M. Ratcliffe of the *Pittsburgh Courier*. Howard almost certainly called Hurley, Evers, and Moore as well. All day and all night that Monday, while the court selected a jury, Howard and his crew looked for the witnesses who could confirm Young's story and Hicks's suspicions. The four they found agreed to come to Howard's house the next evening to relate what they had seen. These evidential gold mines promised not only to provide eyewitness testimony linking the defendants to the murder but also to change the legal jurisdiction from Tallahatchie to Sunflower County. Such a move probably provided the best opportunity civil rights advocates had for disrupting the script of acquittals already unfolding in Sumner.

At eight o'clock Monday evening some members of the Mississippi underground conducted a strategy meeting in Mound Bayou to choose a course of action. Aside from Howard, Hurley was present, as were several African American reporters, including Hicks, Booker, and L. Alex Wilson of the *Chicago Defender*. They agreed that white reporters might fare better in dealing with local law enforcement officials and decided to ask John Popham of the *New York Times* and Clark Porteous of the *Memphis Press-Scimitar* to act in concert with them. All the reporters would have to agree to hold off filing any stories until after the Tuesday night meeting with the witnesses.[5]

What happened next was probably an unintentional blunder. When Howard phoned Porteous to come to a meeting, he neglected to tell him to bring only Popham and to tell no one else. Thus when Porteous arrived that night, he not only failed to bring Popham, but he had recruited W. C. Shoemaker and James Featherston of the *Jackson* (MS) *Daily News*, the most reactionary segregationist newspaper in Mississippi. What is more, rather than demur or equivocate, Howard declared to the three reporters, "I can produce at least five witnesses at the proper time who will testify that Till was not killed in Tallahatchie County but killed in Sunflower County . . . in the headquarters shed of the Clint Sheridan Plantation which is managed by Leslie Milam, brother of J. W. Milam."

It was, as Howard knew, a bombshell.[6] But rather than hide the witnesses in his compound and produce them at the trial at the moment most damning for the defendants, he showed his cards in advance and neglected to swear the reporters to secrecy. Only after considerable wheedling and a promise that they would be the only white reporters invited to the Tuesday evening meeting with the witnesses did Porteous, Shoemaker, and Featherston promise to keep mum for now.[7]

Tuesday's courtroom drama began at a morning break, when Mamie Bradley "walked quietly but purposefully through the center aisle of the courtroom," wrote Rob Hall of the *Daily Worker*, which was filled with "relatives, friends and neighbors of Roy Bryant and J. W. Milam, the two men charged with the murder of Emmett Louis Till, her child." Walking alongside her were two men she introduced to the press as her father, Wiley Nash "John" Carthan of Detroit, and her cousin Rayfield Mooty.[8] Fifty or sixty reporters immediately surrounded her, and photographers leaped over chairs and stood on the black press table to get pictures of her. Sheriff Strider pushed his 270-pound frame through the crowd and handed her a subpoena to be a witness, stating, "You are now in the state of Mississippi. You will come under all rules of the state of Mississippi."[9] Judge Swango seated her near the black press at the front of the courtroom and instructed deputies to locate a larger table for that group. "She is a demure woman whose attractiveness was set off by a small black hat with a veil folded back, a black dress with a white collar," Hall wrote. "In the more than 99-degree heat of the courtroom, she fanned herself with a black fan with a red design."[10]

Arriving with Bradley in a heavily armed motorcade from Mound Bayou was Charles Diggs, U.S. congressman from Michigan, whose family was originally from Mississippi. It took him about an hour to gain entrance, however. Diggs had written to Judge Swango and received a reply inviting him to attend the trial.[11] Even so, according to Hicks, Sheriff Strider initially refused to grant Diggs admission to the courthouse, so Diggs sat in his car with the armed guards and asked Hicks to take the judge his business card. Strider told his deputy, "This nigger here"—gesturing toward Hicks—"says there's a nigger outside who says he's a Congressman, and he has corresponded with the judge, and the judge told him to come on down and he would let him in."

The deputy replied, "This guy said a nigger Congressman?"

"That's what this nigger said," Strider responded and waved Hicks into the courthouse. Hicks made his way inside and sent Diggs's card to the judge, who instructed the bailiff to make room for Diggs at the newly enlarged black press table.[12]

"Local people were obviously surprised when white newsmen shook hands with Rep. Diggs and addressed him as 'Mr. Congressman,'" wrote Rob Hall.[13] A reporter for the *Jackson Daily News* opined, "Diggs has about as much business being at the trial as he has being in Congress" and added that his presence "indicates the political ore to be mined from this judicial molehill by cynical vote-seekers."[14]

It took the court well over an hour to select the two remaining jurors. After the mid-morning break, the jury was seated and District Attorney Chatham took the floor to begin his case. First he summoned these witnesses: Deputy Sheriff John Ed Cothran of Leflore County; Dr. L. B. Otken, who examined the body; C. M. Nelson, the undertaker who dispatched the body to Chicago; Deputy Sheriff Garland Melton of Tallahatchie County, who was present when the corpse was taken from the river; Chester Miller, the African American undertaker from Greenwood who initially took the body; Charles Fred Mims, who helped retrieve the body from the water; W. E. Hodges, a commercial fisherman, and his son Robert Hodges, the seventeen-year-old who initially found the body; Moses Wright; and Mamie Bradley. The defense called the same witnesses plus Roy Bryant, J. W. Milam, Carolyn Bryant, Juanita Milam, Eula Lee Bryant, and Sheriff H. C. Strider. Judge Swango called an early lunch recess at 11:15.

The new jury dined on barbecued pork chops at the Delta Inn, while Milam and Bryant enjoyed the air-conditioning at a lunch spot in Webb with Strider. Mamie, Representative Diggs, and the black press assembled at James Griffin's Place, a black joint on Front Street. During the break, defense attorney Sidney Carlton gathered reporters around him and began to paint Till as a menace to white womanhood who brought his fate upon himself. Till entered the store, "propositioned" Carolyn Bryant, and assaulted her, Carlton said. He "mauled her and he tussled her and he made indecent proposals to her, and if that boy had any sense he'd have made the next train to Chicago." Meanwhile, reporter Clark Porteous, acting as an emissary from T. R. M. Howard, approached the district attorney and gave him a statement from Howard that revealed the existence of the new witnesses and that their testimony would change the location of the murder and tie Milam and Bryant directly to the crime.[15]

This news shocked Chatham and prosecutor Robert Smith. Right after lunch, Chatham likewise shocked the court and caught the defense off-guard. Due to a "startling development" in the investigation, he asked for a recess in order to locate several new witnesses. Chatham said that it might require the entire afternoon to run them all down; though he did not say so, Dr. Howard had arranged to meet with the witnesses and Chatham hoped it would not take that long to assemble them, but managing black murder witnesses in rural Mississippi, where their lives were in danger every moment, could get complicated. Breland leaped to his feet and accused the state of stalling. The trial should proceed at once, he insisted. Judge Swango replied with cool courtesy that the state's request seemed perfectly reasonable to him.

The effort to gain more and better testimony for the prosecution launched what Simeon Booker called "Mississippi's first major interracial manhunt" and Murray Kempton described as "hunting through the cotton fields for four Negroes with a strange story to tell." It involved Booker, Howard, the sheriffs of Leflore and Sunflower counties, Clark Porteous of the *Memphis Press-Scimitar*, W. C. Shoemaker and Jim Featherston of the *Jackson Daily News*, James Hicks of the National Negro Press Association, Clotye Murdock of *Ebony*, David Jackson of *Jet*, L. Alex Wilson of the *Chicago Defender*, Amzie Moore, Medgar Evers, and Ruby Hurley of the NAACP, and perhaps a few others.[16]

Both sheriffs first drove to Leslie Milam's place with Howard and searched the floor of the barn for bloodstains. They found none, but it was obvious that someone had cleaned the floor recently; it was newly covered with corn and soybeans. Unfortunately the investigators did not have the resources or the time to perform a more scientific examination.

Sheriff Smith, who had been battling Sheriff Strider ever since Till's body was retrieved from the Tallahatchie River, acknowledged that he had been looking for witnesses for several weeks and joined the hunt with enthusiasm. "These witnesses have a story to tell," he said. "We've got to find them if it takes all night."[17] The teams agreed to reconvene at eight for a meeting with the witnesses. None of the expeditions went easily. Frank Young did not turn up until one in the morning and refused to talk to anyone but Howard, who was not available.

Moore, Evers, and Hurley put on their farmhand disguises, whisked up the black reporter Moses Newsome of the Memphis *Tri-State Defender*, and combed the plantations and swamplands for witnesses, finding three: Willie Reed, eighteen; his grandfather Add Reed; and their neighbor Amanda Bradley, fifty. Their stories generally confirmed the one Frank Young had told. After Howard promised to protect them in the short term and afterward relocate them to Chicago, all three agreed to testify.[18]

A number of things about the Till trial did not conform to the stereotypes of Mississippi justice in 1955. Judge Swango's fair-minded, even-handed conduct from the bench ran counter to what most observers expected. But perhaps nothing was quite so striking as the workings of Howard's Mississippi underground, a collection of NAACP activists, black and white newspaper reporters, and law enforcement officials who scoured the countryside for witnesses. For Howard and the NAACP contingent, the struggle for justice was motive enough. The reporters sought justice, too, perhaps, but also a story. Judge Swango seems to have genuinely wanted a fair and impartial trial, though it was perhaps more a question of honor than outcome. As for the two sheriffs, they knew that these witnesses could shift the trial from Strider's jurisdiction to their respective jurisdictions, but was a desire for justice their motive for joining the search? They may have simply disliked Strider. Or perhaps they just wanted to be able to face themselves in the mirror. Whatever their reasons, this strange, seemingly fearless group swung into action and found the only witnesses for the prosecution that could tie Milam and Bryant to the scene of the crime.

14

"THERE HE IS"

The trial resumed at 9:20 on Wednesday morning. As Moses Wright made his way toward the front of the sweltering courtroom, quiet fell so that you could hear feet shuffling and the low whump, whump, whump of the ceiling fan. It was the third day of the trial. The authorities had brought in a hundred or so cane-bottom chairs in an effort to keep people off the windowsills and away from the faded lime-green walls. If it had been empty with a good breeze, the room would have reached ninety degrees that day; stuffed sweatbox fashion, the temperature likely reached a hundred or more. The *Delta Democrat-Times* called the courtroom "an oven-hot, smoke-filled room that was jammed to the walls with spectators."[1] The two four-blade ceiling fans seemed only to stir the cigarette smoke. Such oppressive heat discouraged movement other than the polyrhythmic batting of several dozen handheld cardboard church fans of the sort common in the South before air-conditioning.[2]

After two days of jury selection and delays, the short, wiry, dark-skinned preacher was the first witness called. That was not the only reason for the rapt attention in the room, however. Moses Wright was a black man called to testify against two white men charged with murder. In Mississippi that constituted an almost suicidal affront to white supremacy. And he had been duly warned.

Neatly dressed in a white shirt, black pants, a thin, dark blue tie with light blue stripes, and white suspenders, Wright settled into the big wooden witness chair, the back of which reached nearly to the top of his head. He tugged nervously at his thick, workingman's fingers that had been clearing fields of cotton. "I wasn't exactly brave and I wasn't scared," he said later. "I just wanted to see justice done."[3]

District Attorney Chatham cast his first witness in the role of the kindly old black retainer, calling him "Uncle Mose" and even "Old Man Mose" throughout his testimony. Very likely Chatham was playing to his jury, knowing undue respect shown a black man, beyond a kindly paternalism, would only hurt his case. But Wright's presence and demeanor—he sat ramrod straight in the wooden chair—commanded attention, and the DA's questions soon cut to the heart of the matter. "Now, Uncle Mose, after you and your family had gone to bed that night, I want you to tell the jury if any person or if one or more persons called at your home that night, and if they did what time was it?"

"About two o'clock," Wright answered. "Well, someone was at the front door, and he was saying, 'Preacher—Preacher.' And then I said, 'Who is it?' And then he said, 'This is Mr. Bryant. I want to talk to you and that boy.'" When he opened the door, though he acknowledged that he could see neither of the men all that well, he recognized J. W. Milam.

The district attorney asked, "You know Mr. Milam, do you?"

"I sure do," replied Wright.

"And what did you see when you opened the door?"

"Well, Mr. Milam was standing there at the door with a pistol in his right hand and he had a flashlight in his left hand."

"Now stop there a minute, Uncle Mose," instructed Chatham. "I want you to point out Mr. Milam if you see him here."

Moses Wright stood up as tall as his five feet three inches would take him, pointed "a knobby finger at J. W. Milam," and said, "There he is," reported the *Greenwood Commonwealth*, a local white newspaper.[4]

The photographer Ernest Withers raised his camera and took a picture of the cotton farmer in his crisp, clean shirt and neat, thin tie, standing straight and pointing at Milam, who shifted nervously in his chair, puffing a small cigar. One of the wire services bought Withers's roll of film on the spot and the photograph, carried by newspapers around the world, became an iconic image of courage.[5]

"And do you see Mr. Bryant in here?" asked Chatham. Wright pointed again.

"Uncle Mose," Chatham continued, "do you see any man in this courtroom now who was with Mr. Milam that night at your house?"

"Yes, sir."

The defense interrupted with an objection, to no avail.

"And will you point that man out, Uncle Mose?" asked Chatham.

"It was Mr. Bryant," said Wright, rotating slightly and pointing at Milam's half-brother again. Bryant betrayed no emotion, but Milam again shifted nervously in his chair.[6] Only somewhat obscured by the passage of time, the significance of what had just occurred, arguably as significant as anything that would transpire in the courtroom, wasn't lost on most of those watching the testimony. No doubt many thought that Wright had pronounced his own death sentence by identifying the two white men who had taken his nephew. He knew the risks as well as anyone. Murray Kempton wrote that Wright then "sat down hard against the chair-back with a lurch which told better than anything else the cost in strength to him of the thing he had done."[7]

The district attorney's next questions carefully took Wright through his account of the kidnapping, ending with his story of standing on the porch for twenty minutes, watching the darkness into which the boy his niece entrusted to his care had disappeared. "Now tell the Court and Jury when was the next time after they took Emmett Till away from your home that you saw him or his body," Chatham finally directed.

Wright replied, "I saw him when he was taken out of the river."[8]

After Wright explained that he had identified his nephew's body in part by the silver ring on his finger, it was the defense counsel C. Sidney Carlton's turn to cross-examine. Carlton and Wright had spoken three days earlier, of course, when Carlton, in an illegal attempt to prevent the moment that had just come and gone, had dropped by Wright's tenant house to advise him that testifying against Milam and Bryant would be very bad luck, to put it mildly. He tried a less subtle approach in court. "Sidney Carlton roared at Moses Wright as if he were the defendant," Kempton wrote, "and every time Carlton raised his voice like the lash of a whip, J. W. Milam would permit himself a cold smile."[9]

Carlton berated Wright as though the older man were changing the facts of his story, which he wasn't, and Wright calmly pointed out that he had not said any such thing. The defense attorney tried to insinuate that there had not been enough light in the cabin for Wright to identify Milam or Bryant, that his identification of them was mere speculation. He tried to fool Wright into testifying that Emmett Till's own initials were on the silver ring rather than his father's. He tried to suggest that since Wright had testified that he could not see the men putting the boy into the car and had not been able to see Till in the car as it pulled away, he could not prove that they had taken the boy. He attempted to shake Wright's confidence in his identification of his nephew's body by the river. Alternating between accusatory and indignant theatrics, however, Carlton never tripped up Wright, who stuck to answers like "That's right" and "I didn't say it" and "I sure did." When Carlton's frustrated questions became repetitive, Judge Swango sustained the prosecution's objections, the defense lawyer gave in, and the judge announced a twenty-five-minute recess.[10]

Moses Wright did not go into hiding for the rest of the trial. He did not follow his wife and leave immediately for Chicago. Every day of the trial he could be seen around the courthouse, wearing blue pants and a crisp white shirt, his pink-banded hat tilted back on his head. Wright seemed transfigured by his bravery on the witness stand. "He walked through the Negro section of the lawn," wrote Dan Wakefield for the *Nation*, "with his hands in his pockets and his chin held up with the air of a man who has done what there was to do and could never be touched by doubt that he should have done anything less."[11]

When court resumed, defense attorney Breland said, "If the Court please, the Clerk has just handed Defense Counsel a list of additional witnesses which the Clerk states he has subpoenaed both for the state and defense. We now move the Court that the defendants' counsel have the opportunity of examining these witnesses in the witness room before they are offered as witnesses by the state. The names of these witnesses," he continued, "are as follows: Amandy Bradley, Walter Billingsley, [Add] Reed, Willie Reed, Frank Young, and C. A. Strickland." Except for Strickland, the police photographer, these were the local African Americans discovered by the Mississippi underground, and the defense had no idea what they might tell the court. The judge acceded to the request and let the state call its next witness, Chester Miller, the African American undertaker from Greenwood who had taken Till's body from the riverside to his funeral home.[12]

On August 31, Miller testified, Deputy Sheriff John Ed Cothran called him to pick up a body in Tallahatchie County. He took one of his helpers and found the body lying in a boat beside the river. Tallahatchie County law enforcement officers suggested that Miller remove a silver ring from the finger of the deceased for purposes of identification. "I laid it on the floorboard of the ambulance," he said. Before he could load the body into the ambulance he had to detach the barbed wire that lashed a heavy gin fan around the neck. "It was well-wrapped," he testified.

He had then asked Moses Wright to identify the body, which Wright did. In the courtroom, Special Prosecutor Robert Smith asked Miller, "In your opinion, was the body that was put in your ambulance, was it possible for someone who had known the person well in their lifetime to have identified that person?"

Before the defense could shout their objection, Miller replied, "Yes, sir," but Judge Swango sustained the objection and asked the jury to disregard the answer.

Taking another tack, Smith asked Miller to describe the body. "Well, it looked to be about five foot four or five inches in height," the undertaker said. "Weight between one hundred and fifty or sixty pounds. And it looked to be that of a colored person."

"Could you tell whether it was the body of a young person, or middle age or an old person?" Smith inquired.

"It looked like it was the body of a young person."

Miller told the prosecutor there was a hole in the head that "looked like a bullet hole." The defense again objected successfully, though Miller managed to say on the record that it was a half-inch hole just above the right ear. Smith asked about the other side of the head. "Well," said Miller, "it was crushed on the other side. You couldn't tell too much, it was crushed so. And it was all cut up and gashed across the top there."

"Would you state whether or not the wounds described here were sufficient to cause his death?" Smith asked.

Defense attorney Breland broke in: "We object to that, Your Honor. He is no expert to that. And the jury knows as much as he does about that. I think that is within the province of the jury."

"I am going to let the witness answer the question," Swango responded.

Smith asked the question again and Miller said, "Yes, sir."

"I believe I asked you this, but I am not sure," Smith continued. "You testified that there was some barbed wire in the boat. But did I ask you whether or not the barbed wire was on the person of the deceased?"

"Yes, sir," replied Miller. "Around the neck."

For anyone striving to believe otherwise, it was growing harder and harder to evade the conclusion that a murder had been committed. But on cross-examination Breland went after the ghastly description of the injuries to the body. "Now, what you saw about the condition of that man as to his head, you couldn't tell whether it was caused before or after his death, could you?"

"No, sir," said Miller.

"And you couldn't tell whether it was caused in a car accident or some otherwise, could you?"

Miller responded, "No, sir."

The court then recessed for lunch.[13]

The defense was offering the jury a thin veil behind which they could pretend to believe that the savage injuries might have occurred in an unknown manner to an unknown person and that the body had then been dumped in the river, recovered, and presented in a gruesome, politically inspired hoax as the body of Emmett Till. It remained to be seen whether the twelve jurors would take up that veil, or even needed to.

After the lunch break Sheriff Strider stared at the black reporters gathering their things for the afternoon session. Some of them had been part of the Mississippi underground that had convinced the additional witnesses to come forward. All of them would be reporting the trial's proceedings to the world. "Hello, niggers," he said.[14]

Robert Hodges, the seventeen-year-old fisherman who found the body snagged in the muddy water of the Tallahatchie, opened the afternoon testimony, followed by B. L. Mims, whose motorboat had pulled the body to shore. In each case the body began to appear more and more indisputably that of a murder victim.[15]

Then Sheriff George Smith, at least an honorary member of that Mississippi underground that had located the new witnesses, took the stand and testified that he had found Roy Bryant sleeping at the store in Money at about two that Sunday afternoon. He'd had a long talk with Bryant in the front seat of the squad car. "I asked him why did he go down there and get that little nigger boy, and he said that he went down there and got him to let his wife see him to identify him, and then he said that she said it wasn't the right one, and then he said that he turned him loose."

Because Sheriff Smith's testimony conveyed Bryant's admission that he had kidnapped Emmett Till the defense moved quickly to frame the exchange as a private, confidential conversation between two old friends that Bryant never understood was part of a murder investigation. This was before the days of *Miranda*, the Supreme Court decision, later familiar to all fans of crime dramas, that decrees suspects of a crime be advised of their right to remain silent and that everything they say can and will be used against them in a court of law. Consequently all the defense could do was insinuate that the interrogation was dishonest, that a uniformed officer interviewing a friend without declaring the risks of admitting to kidnapping and perhaps worse was an act of deceit to which a reasonable jury should take affront.[16]

Next the prosecution called Deputy Sheriff John Ed Cothran of Leflore County. Cothran had been present when the body came out of the water. He had also been the one to arrest Milam and had talked with him about what had happened to Emmett Till. "I asked him if they went out there and got that boy," Cothran testified. "I didn't call anyone by name. I just asked him if they had went out and got that boy. And then he said, yes, they had got the boy and then turned him loose at the store afterwards, Mr. Bryant's store."

Carlton quickly objected to this testimony and Swango sent the jury out of the room. Carlton suggested that Cothran's status as "a good friend of that entire family" made it all the more imperative that he should have informed Milam of the risks of his admission to breaking the law. "And we have a further objection to this witness's testimony at this time on the grounds that there has been no showing whatsoever in the record that the body taken from the Tallahatchie River and alleged to be that of Emmett Till, that the death was caused by any criminal agency whatsoever." The judge overruled Carlton's objections and brought the jury back in.

District Attorney Chatham then asked Cothran, in the presence of the jury, if he'd ever "had occasion to investigate the murder or disappearance of Emmett Till" and had ever talked to J. W. Milam about that. Cothran answered both questions in the affirmative and added that he'd talked to Milam in the Leflore County Jail the day he'd arrested him. Cothran assured the court that he had not promised Milam anything nor threatened him in any way. "I asked him if they went out there and got that little boy, and if they had done something with him. And he said they had brought him up to that store and turned him loose." This second instance of the defendants' admission to kidnapping went without further comment.[17]

Under questioning by Chatham, Cothran described the scene at the riverbank after he'd arrived to find the body in a boat on the shore. Seeking to undermine the defense's implication that there may have been no murder, Chatham asked about the condition of the corpse. "Well, his head was torn up pretty bad. And his left eye was about out, it was all gouged out in there, you know," Cothran answered. "And right up in the top of his head, there was a hole knocked in the front of it there. And then right over his right ear—well, I wouldn't say it was a bullet hole, but some of them said it was."

Breland piped up from the defense table, "We object to what they said it was," and Judge Swango sustained the objection.

"There was a small hole in his head right above the ear," Cothran corrected himself, "over on the right side of his head, over here," gesturing toward the right side of his own head, "and that was all tore up. There was a place knocked in his forehead." On cross-examination Carlton once again tried to suggest that Cothran's friendship with the accused invalidated his testimony and that the damage to Till's head could have occurred after the body was already in the river. With that the court recessed until ten o'clock Thursday morning. That thin veil the defense was offering the jury now required them to ignore the fact that both defendants had admitted to abducting the boy at two in the morning.[18]

After the session ended on Wednesday, members of the United Packinghouse Workers of America delegation from Louisiana found Moses Wright standing alone outside the courtroom. One of them, Frank Brown, "asked the old man where he found the courage to testify in the face of probable death. 'Some things are worse than death,' Wright told Brown. 'If a man lives, he must still live with himself.' "[19]

On Thursday morning Emmett Till's mother made her way through the crowd and up to the witness stand. Mamie Bradley "was a composed and well-spoken witness," wrote John Popham of the *New York Times*. "She wore a black dress with a white collar and a red sash. She is a pretty brunette." Murray Kempton noted that she "wore a black bolero and a printed dress with a small black hat and a piece of veil and she was very different from the cotton patch cropper who is the ordinary Negro witness in a Mississippi courtroom." A columnist for the *Greenwood Common-wealth* wrote, "The fashionably dressed 33-year-old negro woman had an air of confidence and determination. . . . Her answers were direct and to the point, using good English and speaking in a highly audible tone. At only one point did she display any emotion. This occurred when she was shown a photograph of the body. After looking at the picture, Mamie sobbed, took off her glasses and wiped tears from her eyes." Throughout, the Greenwood paper attested, she appeared dignified, intelligent, sympathetic, and respectable.[20]

Mamie put on her glasses as Special Prosecutor Robert Smith began the examination. After establishing that Emmett's father had been killed in World War II, "in the European theater," Smith asked about the boy's trip from Chicago, when Mamie had learned that he was missing in Mississippi, and where she had first seen her boy's body when it returned from Chicago in a box. It was at the A. A. Rayner Funeral Home in Chicago, she replied.

> The first time I saw it, it was still in the casket. I saw it later on after it was removed from the casket and placed on a slab. . . . I positively identified the body in the casket and later on when it was on the slab as being that of my son, Emmett Till. . . . I looked at the face very carefully. I looked at the ears, and the forehead, and the hairline, and also the hair, and I looked at the nose and the lips and chin. I just looked at it all very thoroughly. And I was able to find out that it was my boy. And I knew definitely that it was my boy beyond a shadow of a doubt.

"I now hand you a ring, Mamie," Smith intoned, "that has engraved on it 'May 25, 1943,' with the initials 'L.T.,' and I ask you if that was among the effects that were sent to you which were purported to be the effects of your dead husband?"[21] Smith called her by her first name, as was customary when a white person addressed an African American in 1950s Mississippi. By contrast, she always referred to Smith as "Sir." Knowing the ways of the South as she did, Mamie Bradley accepted this disrespectful treatment with grace. But a performance perhaps necessary for a white jury in Mississippi played differently in the wider mid-twentieth-century world. The *Washington Afro-American*'s headline, for instance, was "Mother Insulted on Witness Stand."[22]

"Yes, Sir," she replied. "I kept the ring in a jewelry box but it was much too large for the boy to wear. But since his twelfth birthday, he has worn it occasionally with the aid of scotch tape or string." He wore it when he left home for Mississippi, she said. "And I remember that I casually remarked to him, 'Gee, you are getting to be quite a grown man.' "

"And that was the ring he had when he came down to Mississippi?" asked Smith.

"Yes, Sir."

Finally Smith asked her to look at a police photograph of her son's body taken in the funeral home in Greenwood. "And I hand you that picture and ask you if this is a picture of your son, Emmett Till?"[23]

Taking the photograph in her fingers, Mamie bowed her head and wept, rocking slowly from side to side. Then, pulling off her glasses, she wiped her eyes and replied, "Yes, Sir."[24]

When defense attorney Breland cross-examined the witness he, too, called her by her first name. "Mamie," he asked, "where were you born?"

"I was born in Webb, Mississippi."

"That is a little town just two miles south of here, is that right?"

"I can't tell you the location."

"When did you leave Mississippi?"

"At the age of two."

"Then you have just been told that you were born in Webb, Mississippi? You don't remember, is that right?"

"Yes, Sir."

"When you can first remember, where were you living?"

"In Argo, Illinois."

"How far from that is Chicago?"

"Approximately twelve miles."[25]

With this new line of inquiry, the sons of Mississippi were no longer on trial. Now it was Chicago, Chicago that sent its swaggering black males south, Chicago that poured undue scorn hot and fast on the Magnolia State, Chicago that encouraged Mamie Bradley to declare that the entire state of Mississippi would have to answer for this murder, Chicago that was on trial. Had not Mississippi suffered enough at the hands of uppity northern blacks and obstreperous, meddling Yankees? The trouble did not start down here, was the defiant implication.

Was Emmett ever in trouble in Chicago? Breland wanted to know. Mamie stated that he never had been, but that didn't matter; the question was intended to answer itself. For some percentage of the people in that courthouse Emmett's being a black boy from Chicago answered the question, just as his winding up butchered and discarded in a river was explained clearly enough to some percentage of the country by the notion that a black boy had misstepped and thereby had some responsibility for what had been done to him.

Breland then pivoted, training his questions to discredit Mamie herself. "Did you have any life insurance on him?" She did. "How much did you have?"

"About four hundred dollars straight life. I had a ten-cent policy and a fifteen-cent policy, two weekly policies, and they equaled four hundred dollars."

"To whom were those policies made payable? Who was the beneficiary in those policies?" Another fig leaf was being handed the jury.

"I was the beneficiary on one and my Mother on the other," Mamie answered.

"Have you tried to collect on those policies?"

"I have been waiting to receive a death certificate."

Suggesting that she was trying to capitalize on her tragedy was only part of Breland's intent. He was also implying that since she had not tried to collect on the life insurance policy, since she had no death certificate, perhaps there had been no death. Here Sheriff Strider's theory of the case, that it was a put-up job by the NAACP and the corpse was not even Till's, eased into view.

"Now, Mamie, what newspapers do you subscribe to in Chicago? Do you read the *Chicago Defender*?"

Knowing the sympathies of the jury, the prosecution objected: "If the Court please, I think it is perfectly obvious what he is trying to get at. And I think counsel should be counseled not to ask any more questions like that."

"The objection is sustained," Judge Swango pronounced. "Now, will you gentlemen of the jury step back into the jury room a moment, please?" The twelve Mississippians filed through the door behind the witness stand and closed it behind them.

In the absence of the jury Breland questioned Mamie about her subscription to the *Chicago Defender*, which she read every week. "These papers are edited by colored people, is that right?"

"Yes, Sir."

Breland held up a copy of the *Defender*, then handed it to her, asking if she had seen this issue, and more specifically if she had seen the photograph of her son in it. It is of a smiling, utterly boyish Emmett, inescapably fourteen, his dark tie against a white shirt lending him the air of Sunday-best efforts, a hat on his head, the image of a boy hopeful about becoming a man. Did she have a copy of this photograph with her? She had. And when was it made? All of the photographs in all of the newspapers were made on the same day, two days after Christmas in 1954, Mamie told him.

"Did you have several of those photographs made?" She acknowledged that she had. "And did you furnish any of those photographs to members of the press?"

"Yes, Sir."

"And that was for photographic purposes to put in the papers, is that right?" Breland had a number of different papers with him, all of them littered with adorable photographs of her child. She had provided the newspapers with copies. In fact, she had a number of copies with her now, presumably in case other newspapers wanted to feature his story. But there was more.

"Mamie," Breland said momentously, "I hand you a paper, being page 19 of the *Chicago Defender*, on the 17th of September, 1955, which purports to be a photograph of some person. Will you look at that and state whether or not this is also a photograph of Emmett Till or the person who was shipped back to Chicago that you saw at the funeral home there?" It was the grisly photograph of her son's battered, bloated body.

"This is a picture of Emmett Louis Till as I saw it at the funeral home," she said.

"And being the photograph of the same body which you then identified as Emmett Till? And which you now identify as that of Emmett Till, is that right?"

"Yes, Sir."

The judge then asked Breland if he had finished his examination. "I believe we have, Your Honor. And we submit that these are proper at this time."

The defense raised the matter of the *Chicago Defender* and the photographs of Emmett in order to tie Mamie to the national outrage pouring onto Mississippi from the Northern press, particularly the African American press. The imagination was left to fill in the events between the two photographs, which for many Northerners and black people were sufficient for disgust, outrage, and scorn. This scorn from outsiders had become very unpopular among white Mississippians, even those who had no sympathy for the killers. The photographs and the world's disgust made it far harder to avoid the fact that whatever had happened to that boy had happened in their state, under their collective watch, and therefore with some degree of their collective culpability. In story after story much of the world's media made sure that point was made. Breland's associating Mamie with the national press coverage marked her as an outsider and focused the hostility and resentment of the jurors and the public on her.

Smith injected, "Your Honor, we think this is highly incompetent, this whole part of the case."

The judge replied:

With reference to that, I believe that the witness testified that the pictures taken—that one of them is a picture of her son that was taken shortly after Christmas, and I believe that the witness testified that it is a true likeness of her son during his lifetime. And she also testified that the picture taken in Chicago after his death portrays a true picture of what she saw there at that time.

Now, the Court is going to admit these pictures in evidence—that is, one picture there that she produced, so that the jury may see the likeness of Emmett Till during his lifetime. And the Court is going to let be introduced in evidence the picture made in Chicago after his death. It will be cut from the paper, and the paper itself will not be any part of the exhibit.

There will be no reference to any newspapers to which this witness may subscribe in Chicago, or any reference to what she may read. And there [will] be no reference or anything said about any newspapers or pictures other than this picture, which she had identified as being a picture of her son taken after his death as she saw it there in Chicago. That picture will be permitted.

Breland informed Judge Swango that he had one more topic that he wanted to broach with the witness, one that might well be objectionable to the defense. The judge urged him to go right ahead and get it out of the way. The jury remained in the jury room. This next line of questioning was an admission of a different sort, but one that for much of the world implicitly defended Bryant and Milam's kidnapping of Till at two o'clock in the morning.

"Did you caution [your son] how to conduct himself and behave himself while he was down here in Mississippi before he left there?" Breland demanded.

"I told him when he was coming down here that he would have to adapt himself to a new way of life. And I told him to be very careful how he spoke and to whom he spoke, and to always remember to say 'Yes, Sir' and 'Yes, Ma'am' at all times."

"And did you direct his attention as to how to act around white people, and how to conduct himself about a white man? And did you caution him in those conversations you had with him not to insult any white women?"

"I didn't specifically say white women. But I said about the white people. And I told him that because, naturally, living in Chicago, he wouldn't know just how to act, maybe."

"Prior to his coming down to Mississippi," Breland pressed, "and prior to his leaving Chicago, while he was living there in Chicago, had he been doing anything to cause you to give him that special instruction?"

"No, Sir. Emmett has never been in any trouble at any time."

"And he has never been in a reform school?"

"No, Sir."

"I believe you live on the south side in Chicago, is that right? And that is the part of Chicago referred to as the black belt, is that right?"

"Yes, Sir."

"And the people in the community, are they all colored people or white people?"

"There are a few white people living there."

"And they have their homes there, is that right?"

"Yes, Sir."

Breland's implication was clear: untutored, swaggering, race-mixing South Side Chicago had gotten what it deserved.

"Is that all?" asked Judge Swango. The defense attorney said that it was.

"Now," said the judge, "the objections to all that testimony will be sustained, and there will be no questions along that line whatsoever."[26] And yet, captured in the trial record, Chicago had been placed on public trial in Judge Swango's courtroom. A veil of a different sort had been handed not to the men sequestered in the jury room but to the state of Mississippi. Mamie took her seat and the jury returned to the courtroom.

15

EVERY LAST ANGLO-
SAXON ONE OF YOU

Across the courtroom, Carolyn Bryant claimed decades later, she watched in awe as Mamie Bradley testified. "I had all these things running through my mind," she recalled. "My husband's going to the penitentiary, maybe for life. I have children to support." In her memory, however, her fears did not squelch her astonishment at the African American mother across the room. She could not stop thinking about her. "Here is this woman whose child has been brutalized, just brutalized every kind of way—how could she stand it? I don't know how she went through the trial the way she did."[1]

One answer might be that no African American took the stand for the prosecution without having first thought deeply about what doing so was going to ask of them then and thereafter. In unique ways each had already wrestled with the question of how they would live with the consequences of their testimony. Mamie had decided beforehand what her life would mean from then on. The rest of the black witnesses had already made arrangements to leave Mississippi, probably forever, and move to Chicago—including the next witness, just four years older than Emmett Till, who by agreeing to testify was saying goodbye to his home, his friends, his church, and everything he had grown up around.[2]

Willie Reed was one of the witnesses unearthed by Howard's Mississippi underground as they scoured the cotton farms. He was an eighteen-year-old who lived on the M.P. Sheridan place, a large farm in Sunflower County managed by Leslie Milam. His testimony would tie J. W. Milam to the site of the murder; with it the prosecution shifted from trying to inspire sympathy to offering eyewitness evidence of the crime. Like Moses Wright, Reed was asked to point out Milam in the courtroom. Like Moses Wright, he provided another icon of courage, knowing, as did most in that courtroom, that he would have to move, perhaps change his name, live somewhere else for the rest of his life. He surely also imagined that doing all of this might not be enough, that his life might be taken anyway; he was testifying, after all, against two white men in the murder of a black boy. Nevertheless, when asked to identify the killers, he did not hesitate.

"He is sitting right over there," said Reed, pointing at the bald-headed bear of a man at the defense table. Prosecutor Smith asked Willie, for it was always "Willie" in court and never "Mr. Reed," if he had seen Milam on Sunday, July 28. "I seen him—when I seen him he was coming to the well. . . . The well from the barn on Mr. Milam's place."

Reed had left his grandfather's house early that morning, between six and seven, headed for a nearby store. From there, on his way to his morning's work, he went by Leslie Milam's barn. A truck passed him, a green and white Chevrolet pickup, the top white, the body green. It was full of people. "Well, when the truck passed by me I seen four white mens in the cab and three colored mens in the back. And I seen somebody sitting down in the truck back there. . . . I seen another colored boy." They were sitting on the sides of the truck, Reed said, and had their backs to him.

"Well," he continued, speaking so softly that many in the courtroom could barely hear him, "when I looked at this paper, I was sure—well, I had seen it, and it seemed like I had seen this boy somewhere before. And I looked at it and tried to remember, and then it come back to my memory that this was the same one I had seen in the paper."

"And that was Emmett Till?" asked Smith.

"I don't know if that was him, but the picture favored him," replied Reed, who added that he had walked on past the barn.

"And what did you hear?" inquired Smith.

"It was like somebody whipping somebody."

"We object to that," Breland snapped.

"The objection is sustained," Judge Swango responded.

Smith handed Reed a photograph of Emmett Till. "Now I ask you to look at that picture and I ask you . . . does that or does that not resemble the person you saw sitting there in the back of the truck on that particular day?"

Again Breland objected and again he was sustained.

Smith tried another tack: "Have you ever seen that boy before?"

"It is a picture of the boy I saw on the back of the truck."

"Now, later on in the morning, did you see J. W. Milam out there?"

"Well, when I passed by he came out by the barn to the well."

"Will you state whether he had anything unusual on or about his person?"

"He had on a pistol," said Reed. "He had it on his belt."

"And what did Mr. J. W. Milam do when you saw him?"

"He just came to the well and got a drink of water. Then he went back into the barn."

"Did you see or hear anything as you passed the barn?"

"I heard somebody hollering, and I heard some licks like somebody was whipping somebody."

"What was that person hollering?"

"He was just hollering, 'Oh.'"

"Was it just one lick you heard, or was it two, or were there several licks?"

"There was a whole lot of them."[3]

Silence fell over the room, the Baltimore *Afro-American* reported. "There was no laughter in the courtroom then. Beer drinking dropped to a bare minimum. Bryant and Milam looked a trifle pale and the defense counsel—all five of them—looked worried."[4] The number of facts the jurors were expected to disregard had increased considerably.

Reed told of walking a little farther down the road and stopping at Mandy Bradley's house and talking with her. "And after you left Mandy's house the first time where did you go?" Smith asked Reed.

"I came to the well. . . . I came to get her a bucket of water. . . . I could still hear somebody hollering." Taking Bradley her water, Reed walked on to the store and went home to get dressed for Sunday school. On his way back the truck was gone, the barn quiet.

The defense made two motions to strike all of Reed's testimony, but each time Judge Swango refused. This was the most damning testimony so far, tying Milam directly to the killing of Emmett Till and showing clear evidence of murder. It also changed the place where the murder had been committed from Tallahatchie County to Sunflower County, which could have implications for jurisdiction. Buttressing Reed's testimony were two subsequent witnesses, Mandy Bradley and Reed's grandfather Add Reed. Both of them confirmed the eighteen-year-old's account of that brutal morning. At 1:15 on only the second day of the trial, the state rested its case.[5]

James Hicks, who had done so much to help locate the witnesses and had discovered the story of the witnesses still hidden in Sheriff Strider's jail, was baffled. The closing of the prosecution's case seemed premature; at the least it left several witnesses still to be heard from. Hicks was not alone; the Mississippi underground and the black and white reporters who had helped round up the new witnesses also were surprised the others were not called. Frank Young, whose midnight visit to Dr. Howard's place had set the underground to action, was believed to have important evidence in the case. But Young, reportedly seen outside the courthouse that morning, had disappeared. Later some would fault the prosecution for not managing the witnesses better, though it is hard to imagine that Young's testimony would have been decisive, since Reed told essentially the same story and his grandfather and neighbor confirmed it.[6] All five of the defense lawyers later acknowledged to an interviewer that the state had presented "sufficient evidence to convict" Milam and Bryant. The defense now needed to offer jurors committed to acquittal some plausible pretext for their votes.[7]

The first thing the defense did when the state rested its case was to move that the court exclude all evidence offered by the state and issue a directed verdict of not guilty for both defendants. Judge Swango dismissed the motion out of hand. The second thing the defense did was to call to the witness stand Mrs. Roy Bryant. It was time to play the old song of the Bruised Southern Lily and the Black Beast Rapist. Somewhere, perhaps while her husband's family kept her hidden from the world, hidden even from her own family, Mrs. Roy Bryant seemed to have learned all of the verses.

Carolyn took the stand, swore to tell the truth, the whole truth, and nothing but the truth, and answered Carlton's questions about her name, weight, height, marital status, and the like. "Now, Mrs. Bryant, I direct your attention to Wednesday night, on the 24th day of August. On that evening, who was in the store with you?"

Special Prosecutor Smith broke in: "If the Court please, we object to anything that happened on Wednesday evening unless it is connected up." Breland interjected for the defense that they intended to connect the testimony to existing testimony. Judge Swango retired the jury so they could not hear the discussion or Carolyn's testimony, if any.

To anyone still committed to reading the Till trial as an exploration of facts and justice, this is where things take an odd and revealing turn. The jury had been given powerful prosecutorial evidence that Bryant and Milam kidnapped and murdered Emmett Till. Rather than refute that evidence, the defense now wanted to tell the jury *why* Milam and Bryant had every reason to do so: because that black boy had tried to rape this white Southern woman. There was the oddity: the defense wanted to admit evidence that would further damn their clients, and the prosecution wanted to stop the defense from explaining why their clients were guilty.

So here is another shard of truth, which we must accept if we are to make sense of the trial: faith in our courts and our laws, in the statement chiseled above the columns of the U.S. Supreme Court building—"Equal Justice Under Law"—can obscure the obvious, particularly with the passage of time. There was no equal justice, no universal protection of law in the Mississippi Delta, certainly not in 1955. If the real question was whether or not Milam and Bryant had committed murder, wouldn't each team of attorneys have approached the trial differently? Of course. So why didn't they? The obvious answer is that every lawyer in that courthouse knew that a jury of white male farmers from Tallahatchie County would hear a story about a black boy and a white woman and approve of that boy's murder. The contradiction of a defense team strategizing to introduce a motive for the crime they professed their clients did not commit provided glaring evidence, if any were needed, that the trial had never been about justice.

Fifty years later Carolyn summoned her courage to tell me that her testimony had not been true, even though she didn't remember what *was* true, but that nothing Emmett Till did could ever justify what had happened to him. But in 1955 she provided the court and the case with a billboard: *My kinsmen killed Emmett Till because he had it coming.*

Carolyn knew then, as she would admit much later, that her testimony was a lie. If Till had deserved what happened to him, then why did she hide the incident at the store from her husband and brother-in-law, if that is what she did? If Till had done something terrible, something he ought to have been brutally punished for, then why had she been reluctant to identify him? If he had laid his hands on her, then why didn't she tell her lawyer so only a few days after it happened? Why did her husband and brother-in-law persistently refer to Till's alleged crime as "smart talk" and "ugly remarks"? Is it plausible that Till put his hands on Carolyn and yet his assailants referred only to his verbal transgressions, never uttering a word about something approaching a rape attempt?

But that day in the Sumner courthouse, the jury didn't even need to hear her testimony. After a brief discussion of legal points, the question before Judge Swango boiled down to this: the unchallenged testimony in court had heretofore been that Milam and Bryant came to get "the boy that did the talking over at Money," and now the defense wanted to fill in the details of just what that "talking" had concerned. Except whatever may have occurred in the store at Money clearly did not alter the facts of a kidnapping and murder that occurred several days later at the instigation of the defendants. What should not have been a matter of courage—ruling appropriately on a matter of law—became one in that Jim Crow court-room: Judge Swango ruled that the jury was not going to hear Carolyn Bryant's testimony.

The world, however, would hear it. After Swango's ruling, Breland said, "We wish to develop the testimony for the sake of the record."[8] Why? Here is yet another shard of truth: Breland and his colleagues knew it was next to certain that the jury would hear about Carolyn's testimony within a few hours. Indeed they knew that the jury had taken their seats in the jury box with some understanding of what Carolyn was going to say, what the combination of a dead black boy and an affronted white Southern woman implied without anyone saying anything further, their respective roles firmly established over centuries. And as the jurors filed out of the courtroom, Carolyn Bryant was at the witness stand, having come to play a part that each of them knew by heart. So, "for the sake of the record," Carolyn Bryant testified.

"This nigger man came in the store and he stopped there at the candy counter." The counter was at the front of the store, on the left. "I asked him what he wanted," and he ordered some candy. "I got it and put it on top of the candy case. I held out my hand for his money. He caught my hand." She demonstrated his grip.

"By what you have shown us," said Carlton, who was handling the direct examination, "he held your hand by grasping all of the fingers in the palm of his hand, is that it?"

"Yes. I just jerked it loose."

"Just what did he say when he grabbed your hand?"

"He said, 'How about a date, baby?' I turned around and started back to the back of the store. He came on down that way and caught me at the cash register. Well, he put his left hand on my waist and he put his other hand on the other side." At the request of her attorney, she stood up and placed his hands on her body in just the way she said the boy did.

"Did he say anything to you then at the time he grabbed you there by the cash register?"

"He said, 'What's the matter, baby? Can't you take it?'" It was with considerable difficulty that she was finally able to free herself from his hold on her. "He said, 'You needn't be afraid of me.'"

"And did he then use language that you don't use?" Yes, he had done that. "Can you tell the Court just what that word begins with, what letter it begins with?" She shook her head. No, she could not even say the first letter.

"In other words, it is an unprintable word?" It was. "Did he say anything after that one unprintable word?"

"Well, he said, well, '——with white women before.'"

"When you were able to free yourself from him, what did you do then?"

"Then this other nigger came in the store and got him by the arm. And then he told him to come on and let's go."

"Did he leave the store willingly or unwillingly?"

"Unwillingly," she replied. "He had him by the arm and led him out."

"When he went out the door, did he say anything further after he had made these obscene remarks?"

"Yes. He turned around and said, 'Goodbye.'"

There were perhaps eight or nine black boys out front, she said, including her assailant, but she yelled back for Juanita Milam to watch the store and then walked out through this crowd of boys to Juanita's car. She grabbed the pistol from under the front seat. The black boy who had seized her around the waist and uttered unspeakable obscenities was standing by one of the posts on the front porch. And he whistled at her.

"Was it something like this?" Carlton whistled. She nodded yes. "Did you have any white men anywhere around there to protect you that night?"

"No."

"Was your husband out of town?"

"He was in Brownsville. He had carried a load of shrimp there."

"When did you expect him home?"

"I didn't know." That was why Juanita was with her, she explained. "So that I wouldn't be alone."[9]

After Carolyn returned to her seat and the twelve men returned from the jury room, Dr. L. B. Otken testified for the defense. Having justified the murder of Emmett Till to uphold the purity of white Southern women, the defense shifted back to argue that their clients did no such thing. The boy had it coming, in other words, but our clients did not kill him. In fact, although it was his fault if he was dead, he might not even be dead.

A practicing physician in Greenwood, Otken had apparently viewed but did not examine Till's body in "the colored funeral home." Even so the defense regarded him as an "expert witness" because he was a physician who had experience with dead bodies. "This body was badly swollen, badly bloated," Otken said. "The skin and the flesh was beginning to slip. The head was badly mutilated. The right eye was protruding. And the tongue was protruding from the mouth." So far his testimony was no more than any reader of the *Chicago Defender* or *Jet* could have told the court. "I would say it was in an advanced state of decomposition," he added. And then the point: "I don't think you could have identified that body."

"Could a mother have identified that body, in your opinion?" asked Breland.

"I doubt it."

On cross-examination Special Prosecutor Smith asked Otken whether he "could tell if this was the body of a colored person or a white person."

Otken answered that he could not tell, which somewhat begged the question of why the body went straight to "the colored funeral home," death in Mississippi being anything but integrated.

On redirect examination Breland asked Otken if he could tell whether the injuries to the body that he described were present before death or could have been inflicted on the body after death. "I couldn't," answered Otken. In short, it was his studied opinion that the body had been in the water longer than Emmett Till could have been; that it was impossible to identify the body, even as to race; and that the mutilation could have been caused by the body bouncing along the bottom of the river, not evidence of murder.[10]

Sheriff Strider got the last word. He had gone to the river at about 9:15 on August 31, he told the judge, to see the body that people claimed had belonged to Emmett Till. "Well," he said, "it was in pretty bad shape." The skin had "slipped on the entire body." There was a small hole in the head and two or three serious gashes. He estimated that the body had been in the river "at least ten days, if not fifteen," which is to say longer than the body of Till could have been in the water. He could not even swear that the corpse was that of a colored person, a point he reiterated: "At the time it was brought out of the water he was just as white as I am except for a few places." To underscore the point he helpfully added, "If one of my own sons had been missing, I could not swear if it was my own son or not, or anyone else's." He had signed the death certificate, true, but he did not recall whether or not it had Till's name on it. Strider claimed that he'd "had several reports about a negro who disappeared over there at Lambert," but he'd gotten conflicting stories. He'd been unable to investigate further. "I have been tied up here in court."[11]

The following morning the defense called a handful of friends to attest to the fine character and neighborly attributes of J. W. Milam and Roy Bryant. Then the defense rested its case and motioned to exclude all the evidence the state had presented against the defendants and direct the jury to send in a verdict of not guilty. Judge Swango overruled the motions, maintaining that the evidence presented some questions for the jury to answer for themselves. He ordered a fifteen-minute recess, until 10:38, when attorneys for each side would present summations.[12]

Gerald Chatham, the powerfully built district attorney, delivered the opening summation for the prosecution, followed by summations from several members of the defense team, with Robert Smith of the prosecution closing out. Chatham spoke with powerful emotion that seemed to wring sweat from his body; his shirt was dripping wet by the time he finished speaking in the sweltering courtroom.

"By every courtroom standard, the Mississippi born district attorney made a great plea for the dead colored youth," James Hicks wrote, comparing Chatham to a Southern evangelist. "For his numerous moments of brilliant oratory, he brought tears to the eyes not only of those seated at the colored press tables but to some of the white listeners as well."[13] Pounding the table occasionally, Chatham asserted that he was not moved by "the pressure and agitation of organizations outside or inside the state of Mississippi." In other words, he didn't like the NAACP any more than the jury did. Instead, he told them, "I am concerned with what is morally right. To be concerned with anything else will be dangerous to the precepts and traditions of the South."

Chatham, knowing his jury well, hewed to a certain vision of Southern identity. Foremost, a true Southerner would never kill a child. "I was born and bred in the South, and the very worst punishment that should have occurred was to take a razor strap, turn [Emmett Till] over a barrel, and whip him. I've spanked my child and you've spanked yours. The fact remains, gentlemen, that from the time Roy Bryant and J. W. Milam took Emmett Till from the home of Mose Wright he hasn't been seen since." What these two former soldiers did was give a fourteen-year-old boy "a court martial with the death penalty."

"The very first words of the State's testimony were dripping with the blood of Emmett Till. What were those words, gentlemen? They were, 'Preacher, preacher, I want that boy from Chicago, the one that did the talking in Money.' . . . That wasn't an invitation to that card game [as] they claimed."[14]

In addition to navigating notions of Southern loyalty and manners, Chatham knew that he had to offset defense arguments that the body in the river was not Emmett Till's. So he told a story about the disappearance of a beloved family dog. His son came to him one day and said, "Dad, I've found Old Shep," and led him to the badly decomposed body of their dog. "That dog's body was rotting and the meat was falling off its bones, but my little boy pointed to it and said, 'That's Old Shep, Pa. That's old Shep.' My boy didn't need no undertaker or a sheriff to identify his dog. And we don't need them to identify Emmett Till. All we need is someone who loved him and cared for him. If there was one ear left, one hairline, then I say to you that Mamie Bradley was God's given witness to identify him."[15]

"They murdered that boy," Chatham said finally, "and to hide that dastardly, cowardly act, they tied barbed wire to his neck and to a heavy gin fan and dumped him in the river for the turtles and the fish."[16]

When Chatham headed for his seat, Hicks heard Mamie, who was sitting next to him, whisper to herself, "He could not have done any better."[17]

C. Sidney Carlton offered the first summation for the defense. He poured himself a paper cup of water from a green pitcher on the judge's desk as he began to speak and sipped it intermittently.[18] The final curious twist in the testimony of Carolyn Bryant then played out. "Where is the motive?" Carlton asked. The incident at the store was immaterial, he suggested to the jury. The "testimony by Bryant's wife did not implicate Till." Her story of a black man who "molested" and whistled at her did not reflect on the boy, and his clients certainly did not kill him on that account. By this rhetorical feint he managed to inform the jury that Carolyn Bryant had testified that she had been sexually manhandled. The inevitable implication: after hearing what happened to her, any red-blooded white man would have responded the same way.

Moving from the racially visceral to the seemingly rational, Carlton reminded the jury of the "scientific" evidence that the battered, bloated body in question had been in the river much longer than Till's could have been. He then questioned the lighting at the Wright home and whether it was plausible that the old man could have made a positive identification. The most damning evidence—that Roy Bryant had said "This is Mr. Bryant"—the defense attorney sought to transform into a weakness: "Had any of you gone to Mose Wright's house with evil intent, would you have given your name? There's nothing reasonable about the state's case." He began to shout: "If that's identification, if that places these men at that scene, then none of us are safe."[19] Of course Carlton didn't mention that Milam and Bryant had both confessed to the kidnapping. There was "nothing reasonable about the state's theory that Milam and Bryant

171

kidnapped Till in Leflore, drove several miles to a plantation in Sunflower County, then doubled back into Tallahatchie County to dump the body into a river."[20] In short, Milam and Bryant hadn't kidnapped or killed Emmett Till, but they would have been justified in doing so. Here was the cold, unspoken fear running through the white South in the 1950s: If we condemn Milam and Bryant, what else must we condemn? If you vote to acquit these defendants, Carlton closed, "may you feel, in the words of Charles Dickens, that 'tis a far, far better thing you do now than you have ever done."

J. W. Kellum, Sumner's homegrown attorney, spoke briefly just before the lunch recess. He called the jury "a peerage of democracy" and "absolutely the custodians of American civilization." A guilty verdict would be tantamount to "admitting that freedom was lost forever," Kellum said solemnly. "I want you to tell me where under God's shining sun is the land of the free and the home of the brave if you don't turn these boys loose." If you do not, "your forefathers will absolutely turn over in their graves." So began his flight of oratory, some of it borrowed from Mississippi's late governor Paul Johnson:

> I want you to think about the future. When your summons comes to cross the Great Divide, and as you enter your father's house—a home not made by human hands but eternal in the heavens—you can look back to where your father's feet have trod and see your good record written in the sands of time. And, when you go down to your lonely, silent tomb to a sleep that knows no dreams, I want you to hold in the palm of your hand a record of service to God and your fellow man. And the only way you can do that is to turn these boys loose.[21]

"I thought I had heard it all during the testimony," Mamie Bradley wrote, "but those defense lawyers were saving the worst for last."[22]

After the lunch recess John W. Whitten provided the final summation for the defense. He took on the job of returning to Sheriff Strider's theory that the body pulled from the river was not the body of Emmett Till but instead part of an NAACP-sponsored scheme to bring down shame upon the state of Mississippi and drive a wedge between the races. He acknowledged that Milam and Bryant may have abducted the boy, but if they did, they had released him. Perhaps then Moses Wright had taken him to the NAACP, who persuaded Wright to plant that silver ring upon another decaying body so that people would assume that it was Till.[23] No doubt the young man had been whisked back to Chicago or to Detroit, where he also had family. Wrote Mamie, "They practically accused Papa Mose and the NAACP of grave robbing."[24]

"There are people who want to destroy the way of life of the Southern people," Whitten intoned. "There are people in the United States who want to defy the customs of the South" by undermining the time-honored and harmonious relations between the races and by "trying to widen the gap which has appeared between the white and colored people in the United States. Those people would welcome the opportunity to focus national attention on Sumner, Mississippi. They would not be above putting a rotting, stinking corpse in the river in the hope that it would be identified as Emmett Louis Till. And they are not all in Detroit and Chicago," he said pointedly. "They are in Jackson, Vicksburg"—where Medgar Evers organized, where the NAACP had filed school desegregation petitions—"and they are in Mound Bayou, too," an unmistakable reference to Dr. T. R. M. Howard and the RCNL. "And if Moses Wright knows one, he didn't have to go far to find him. And they include some of the most astute students of psychology known anywhere. They include doctors and undertakers and they have ready access to a corpse which could meet their purpose."[25]

But none of the evidence really mattered, Whitten told the jury. "It is within your power," he assured them, "to disregard all the facts, the evidence, and the law, and bring in any decision you like based upon any whim. There is no way anyone can punish you for any decision you make. The last time an attempt was made by a judge to punish a jury for refusing to follow his instructions was in England in the time of Charles II, and this was overruled. . . . You are our hope and confidence to send these defendants back to their families happy."[26] Asking the jury directly to acquit

Milam and Bryant, Whitten expressed his full confidence that "every last Anglo-Saxon one of you has the courage to do it."[27]

With that invocation the defense's perverted passion play performed for the Jim Crow faithful was nearly finished. It was only left to the prosecution to interrogate that script just enough and in just the right way so that at least one Anglo-Saxon juror might find the fortitude to question it. And so Robert Smith, the former FBI agent, rose to give the last summation to the jury.

It was plain that he read the jury much as his colleagues for the defense did: that they were a band of sunburned Mississippi farmers outraged about the outsiders who scorned their state and their way of life. They believed in that way of life as a bulwark against foreign ideologies and disloyal revolutionaries. Preferring to acquit Milam and Bryant, well aware of all the agitation from African Americans since Black Monday, they might be persuaded to seize upon the defense's suggestion that the corpus delicti was part of an NAACP scheme. But where the defense played on Anglo-Saxon unity and racial tradition—surely a shrewd if predictable tack—Smith thought it possible that basic, heart-deep honesty and perhaps a shred of paternalism toward black people might give the jury room to believe the more plausible narrative: that an old black uncle and a terrified but brave young black man were telling the truth. To Smith fell the difficult task of lighting a narrow pathway for these twelve white men to follow that would preserve their dignity while it protested certain aspects of their way of life, a pathway that defined and defended that way of life by damning its worst excesses. He was asking them, and his closing comments suggested that he knew it, to be as courageous as Wright and Reed and Mamie Bradley.

Like the defense, Smith called upon the jury as fellow white Southerners, men dedicated to their common homeland, though clearly he urged a different strategy for preserving a Southern social order based on defending human rights: "Gentlemen, we're on the defensive. Only so long as we can preserve the rights of everyone—white or black—can we keep our way of life. . . . Emmett Till down here in Mississippi was a citizen of the United States; he was entitled to his life and liberty."[28]

Also like his counterparts at the defense table, Smith played upon the fears and resentments sown by outsiders, with their high-flown, profit-driven assaults on the honor of Mississippi. "Outside influences," in fact, want the two murderers freed for their own perfidious purposes. "If they are turned loose, those people will have a fundraising campaign for the next fifteen years."

As flimsy as it was, Smith had to deal with Strider's conspiracy theory about the identity of the body and the unscrupulous schemes of the NAACP. The prosecutor described this theory in terms considerably less admiring than the defense had used and ended by calling it "the most far-fetched [proposition] I've ever heard in a courtroom."[29]

At the end of the day the prosecution team was counting on the testimony of its African American witnesses—Moses Wright, Mamie Bradley, and Willie Reed, in particular. Wright tied both defendants to the kidnapping, and both of them acknowledged as much. He also identified the body, though Mamie's testimony went much further to establish that the body belonged to her son. Reed, only eighteen, proved himself as brave as Wright, pointing straight at Milam when asked to identify the man he saw come out of the equipment shed with a pistol strapped to his waist. His account of these events tied Milam and his truck to the place where the murder occurred and was substantiated by two other witnesses. But all these witnesses were black, and for a white Mississippi jury to weigh their word against Sheriff Strider's and the defense's version of events would have been revolutionary.

So when Smith asked the jury to look beyond race for the truth of the brutal murder of Emmett Till, he did so in terms they could accept: "Old Mose Wright is a good old country Negro, and you know he's not going to tell anybody a lie"; Mamie Bradley spoke with the authority of a mother's love; and Willie Reed was a plainspoken young man of great courage. "I don't know but what Willie Reed has more nerve than I have," Smith added. Then he made his way back to the prosecution's table.[30]

At 2:34 the jury retired to the jury room to decide the verdict. Mamie, sitting at the black press table with Representative Diggs and the contingent of reporters, decided she did not need to wait. "As the jury re-tired," she wrote later, "I measured the looks of the folks in the rear and I turned to Congressman Diggs and the others. 'The jury has retired and it's time for us to retire.'" She and Diggs, taking the still-trembling Reed with them, made their way through the milling crowds to their car and headed back to Mound Bayou.[31]

16

THE VERDICT OF THE WORLD

Milam, Bryant, and their accomplices expected a select audience for their abduction and butchery of a black teenager. They expected that audience to be local rather than regional or national, and it is unlikely they gave world opinion any thought. They assumed any attention given their crimes would be whispered rather than broadcast. They abducted Emmett Till, killed him, and disposed of his body in ways they knew would promote those whispers. With their guns and flashlights and swagger, they knew that every black person in the Delta would soon hear of his disappearance, and their role in it. From the moment they got in the truck to grab the boy who "did the talking" they were determined, as they would soon say publicly, to send a message that would serve as both signpost and pillar of the social order of white supremacy. From the moment they decided to kill him, their act was a lynching, not an assassination or a simple murder.

White authorities were determined to claim otherwise. "This is not a lynching," declared Governor White. "It is straight out murder."[1] Yet despite such official and editorial claims, this was a lynching in the sense that a group of people killed someone and presumed they were acting in service to race, justice, tradition, and widely held values in their community.[2] The lynch mob never intended Emmett Till's killing to be one of the old spectacle lynchings, once common in the South, with the victims burned alive or hanged before an audience of hundreds or even thousands, body parts taken as souvenirs, lynching photographs bought and sold, lurid accounts published in local newspapers.[3] As the twentieth century marched onward, extrajudicial murders conducted for public viewing and participation were less acceptable. But while Carolyn Bryant's kinsmen intended, at least initially, that the details of their torture and killing of Emmett Till would remain their own family secret, they knew neighbors would talk, and they expected them to do so. The decision to take the boy started with storefront rumors, and they intended that his murder would become a matter of local gossip and lore, a badge of honor among the faithful.

A quiet joke went around: "Isn't that just like a nigger to try and swim the Tallahatchie River with a gin fan around his neck?"[4] That kind of local winking and terror were as far as the men who killed Emmett Till expected their murderous handiwork to go. Instead Till's body rose from the dark waters of the Tallahatchie, ended up on worldwide television, and painted his death brightly in the unimaginable global imagination. Mass media and massive protest may have made his murder the most notorious racial incident in the history of the world. White mobs lynched thousands of African Americans—even children occasionally—but it is Emmett Till's blood that indelibly marks a before and after. His lynching, his mother's decision to open the casket to the world, and the trial of Milam and Bryant spun the country, and arguably the world, in a different direction.

When the jurors filed out of the courtroom to begin their deliberations, the crowd surrounding the Tallahatchie County Courthouse thinned, but not in expectation of a drawn-out deliberation. Fat raindrops had begun to bounce on the streets and sidewalks of Sumner.[5] The rain did little to cut the late summer's oppressive heat, and after just eight minutes the twelve jurors sent out a request for Coca-Cola. In the courthouse Milam read the newspaper and rocked on the back legs of his cane-back chair. Roy and Carolyn Bryant looked nervous; she worried that her children might grow up without a father in the house and that she would have no way to support them. But whatever nervous tension was running between Roy and Carolyn, it had dissipated enough that by the time the foreman knocked on the inside of the jury room door after about an hour, J.W. and Roy had already lit big cigars, swaggering with confidence in their exoneration.[6] Mamie Bradley, Moses Wright, and all the other black witnesses had long ago left the courthouse.

As the jurors returned, the sun broke through the clouds outside and a loud murmur arose from the crowd. Judge Swango rapped his gavel and decreed that there would be no demonstrations in his courtroom when the jury announced their verdict, nor were any photographers to take pictures.[7]

The members of the jury looked solemn. Judge Swango asked them to stand: "Gentlemen of the jury, do you have a verdict?" J. A. Shaw, the foreman, answered, "We have," then said, "Not guilty." A shout of celebration went up from the crowd, and the judge demanded quiet. He reminded the jurors that he had instructed them to write down the verdict, and sent them back into the jury room. Even so, spectators in the stairwells rushed downstairs, and the refrain "Not guilty" echoed in shouts through the corridors. When the jurors reemerged, the judge asked Shaw to read the verdict aloud, which he did: "We the jurors find the defendants not guilty." By this time word of the acquittals had reached out into the throngs, now increasingly white, that were gathered outside, and a great commotion arose as they sent up a cheer.[8]

Journalists, photographers, and well-wishers crowded around the Milams and Bryants, shaking hands and slapping backs. Photographers urged the couples to kiss for the cameras, which they did. "I don't remember anything about when the verdict was brought out," Carolyn said, "because all that went through my head was 'Oh, thank God, my children have a father,' and so I don't remember." While the Milams appeared genuinely happy, the Bryants' affection seemed oddly forced, perhaps an early sign of the centrifugal forces that eventually would lead to divorce.[9]

The performance of the trial wasn't over, however; now it was the jury's turn onstage. Having fulfilled their civic duty, the jurors filed through the crowded room, exclamations of "Good work" and "Nice going" trailing behind them.[10] One juror explained to reporters that the jury reached its verdict on the third ballot during the hour-long deliberation. "There were several reasons for the verdict," he said. "But generally everyone reached the conclusion that the body was not definitely identified."[11] The first ballot had three abstentions, the second had two abstentions, and the third was unanimous—not one juror had cast a "guilty" ballot. The hour's delay had been staged, with Sheriff-elect Harry Dogan sending word to the jury to "make it look good" by taking their time. "If we hadn't stopped to drink pop," one juror said later, "it wouldn't have taken that long."[12] In public, however, most stayed on script. Shaw, a farmer from Webb—Mamie Bradley's birthplace—explained to reporters why they had voted to acquit Milam and Bryant: "We had a picture of the body with us in the jury room, and it seemed to us the body was so badly decomposed it could not be identified." But in the atmosphere of victory that suffused the Sumner courthouse, decorum was hard to maintain. Asked whether Mamie Bradley's testimony had impressed the jury at all, Shaw sneered, "If she had tried a little harder, she might have got a tear."[13]

Hugh Whitaker, a graduate student from the Sumner area whose father had worked the trial as a law officer, returned half a dozen years later and interviewed nine of the twelve jurors. He found that not one had ever doubted that Milam and Bryant had killed Emmett Till, and only one had even entertained Sheriff Strider's suggestion that the corpse might not be Till's. Nobody had based his vote to acquit on "outside interference" by the NAACP or the flood of reporters and media coverage. All of the jurors Whitaker interviewed agreed that the sole reason they had voted "not guilty" was because a black boy had insulted a white woman, and therefore her kinsmen could not be blamed for killing him.[14]

• • •

Mamie Bradley, Willie Reed, and Charles Diggs were on the highway bound for Memphis, where they would catch northbound airplanes. As promised, Diggs had bought a plane ticket to Chicago for Reed, who had left the Delta only once in his life, and then only to go as far as Memphis. They traveled quietly until the news came. "We were on the road about fifteen minutes out of Mound Bayou," Mamie wrote, "when it was announced on the radio. There was such jubilation. The radio reporter," who appeared to be broadcasting from the courthouse square, "sounded like he was doing the countdown for a new year. You could hear the celebration in the background. It was like the Fourth of July."[15]

Whatever hopes any in that car had maintained during the trial, none of them was surprised at the verdict. Nor was Reverend Wright, his cotton harvested and his duty done. The most thoughtful observers awaited not the jubilation of Sumner but the verdict of world opinion.

The cluster of cameras outside the courthouse should have reminded the men and women of Sumner that much of the planet was not only watching but judging them, judging Mississippi, and judging the United States. That was a realization that would fully descend upon them only over the ensuing months. For now they felt real pride. No doubt the jurors and many local observers would have seconded the secretary of the Citizens' Council, who told Homer Bigart of the *New York Herald Tribune*, "Sir, this is not the United States. This is Sunflower County, Mississippi."[16] Others cheerfully asserted that the trial exonerated Mississippi of the slurs slung at her. It had been a fair trial and a proper verdict, proving the state could handle its own affairs, thank you very much.

The journalists who covered the trial were more circumspect; none of them was cheerful, although most considered it a fair trial but for the verdict. "No prosecutors in the United States could have worked harder or longer for a conviction than did District Attorney Gerald Chatham and Special Prosecutor Robert B. Smith, both native white sons of their state," wrote James Hicks, who had come to Sumner expecting "Mississippi white man's justice" and nothing more. "And no judge, whether on the Supreme Court bench or the rickety rocking chair of the bench at Sumner could have been more painstaking and eminently fair in the conduct of the trial than Judge Curtis M. Swango of Sardis, Miss." Unfortunately, lamented Hicks, white Mississippi and the jurors held tightly to their age-old blind spot of racial prejudice, "which prevents them from seeing and thinking straight when they look upon a black face."[17]

With varying degrees of interest and drawing a wide array of lessons, white America wrestled with what had happened in Judge Swango's courtroom. Almost all of them agreed to knowing what the jury had known, that Bryant and Milam had participated in killing Emmett Till, and for the oldest of reasons white Southern men sometimes killed African Americans: for an unacceptable sexual affront to their sensibilities and status. Not mere prejudice but the inbred fear of the Black Beast Rapist called the tune in Sumner's courtroom, several reporters noted. Max Lerner of the *New York Post* wrote, "On the sanctity of white womanhood, a Mississippi jury is only a vehicle for expressing the mass fear and hatred of the Negro."[18] The accusation that Emmett Till had attacked Carolyn Bryant boiled the blood of white spectators. Jurors had almost certainly heard rumors of "the talk" that had sent Bryant and Milam to kidnap the boy, and their interpretation of that talk was steeped in centuries of fearful myths. In his incendiary summation defense attorney Sidney Carlton stated that Till "molested" Carolyn Bryant as if it were a fact established by the proceedings. Bill Sorrells, who covered the trial for the *Memphis Commercial Appeal*, observed that the defense "was built on emotion and Mrs. Bryant was the key."[19] The defiant editors of the *Greenwood Morning Star* seemed to confirm this analysis; Mississippi's misfortune, they wrote, was partly because the trial had become known as the "wolf whistle case" or the "Till murder case," when all along it should have been called "this rape attempt case."[20]

White Mississippians' reaction to the legal process ranged from the resentful to the surreal. Hodding Carter Jr. at the *Delta Democrat-Times*, a Pulitzer winner and a moderate who had earned the hatred of many white Mississippians, blamed the laxity of law enforcement for the acquittals and blamed the NAACP for that laxity. Had local officials not been put on the defensive, he opined, they "might otherwise have made an honest effort to do more than what resulted in an effective cover-up." Carter sounded almost like the reactionary editor of the *Jackson Daily News* as he blasted the NAACP for "blanket accusations of decent people, their studied needling of the citizens who had to decide a matter of local justice, and their indifference to truth in favor of propaganda-making."[21]

The editors of the *Jackson Daily News* agreed with Carter that the evidence had been insufficient, but they felt none of his disapproval of the outcome: "The cold hard fact concerning the acquittal in Tallahatchie County of the two alleged slayers of a black youth from Chicago is that the prosecution failed to prove its case." The *Memphis Commercial Appeal* affirmed that "evidence necessary for convicting on a murder charge was lacking."[22] A great many white Mississippians responded with pride that the trial had been a paragon of fair play that demonstrated the Magnolia State's critics wrong. "Mississippi people rose to the occasion and proved to the world," declared the *Greenwood Morning Star*, "that this is a place where justice in the courts is given to all races, religions and classes."[23]

Carter became a kind of national expert on the occasional insanity in his adopted home state. Outside Mississippi he spoke of the terrible injustice of the Till affair, but at home he often seemed to consider the blot on Mississippi's good name to be the real tragedy. Attempting to speak for enlightened Southern opinion, he wrote, "It was not the jury that was derelict in its duty, despite the logical conclusions it might have made concerning whose body was most likely found in the Tallahatchie River, and who most likely put it there, but rather the criticism must fall upon the law officials who attempted in such small measure to seek out evidence and to locate witnesses to firmly establish whoever was or was not guilty." He chose to blame the investigation that had produced such weak evidence, not the jury that had assessed it.[24]

In a widely cussed and discussed piece in the *Saturday Evening Post* titled "Racial Crisis in the Deep South," Carter described Mississippi as the most stubborn Southern state in its resistance to integration and called the recent murders of George Lee, Lamar Smith, and Emmett Till "symp-tomatic." Whites considered the NAACP "the fountainhead of all evil and woe," and the factual nature of most of the NAACP's "bill of particu-lars . . . doesn't help make its accusations any more acceptable." He ad-mitted, "The hatred that is concentrated upon the NAACP surpasses in its intensity any emotional reaction that I have witnessed in my Southern life-time." This reflected the NAACP's demands for voting rights and school integration as much as it did their protests over the Till case. Carter also raised the sex bugaboo, describing Till as "sexually offensive" and stating that "sexual alarm on the part of white men may explain the failure to convict the men accused of the slaying of Emmett Till."[25]

Carter often stood on an increasingly precarious middle ground; he was an early if sometimes wavering articulator of ideas that would eventu-ally herald a different sort of South, when the civil rights movement rose to create it, only small thanks to people like him. In this instance, however, he ran hard up against a butchered fourteen-year-old. Carter tried, but there was scarcely a "Southern moderate" place to stand. Many white Mississippians, particularly those running the state, considered him a trai-tor. More liberal critics took shots at him for defending the indefensible. Roi Ottley of the *Chicago Defender* dismissed editor Carter of "pathetic Mississippi" as "a victim of the South's pernicious folkways," a pitiful apologist who was "attempting to smooth over the facts."[26]

Other white Mississippians seemed to believe that communist agents were responsible for promoting racial division. "Could there be any doubt that this Mississippi murder—from the weeks or months before young Till made his visit to the South—was communist-inspired, directed and executed?" wrote a woman from Memphis. "Did not some communist agent or agents murder the young Negro after the white men turned him loose?"[27] A well-dressed local woman in Sumner told the television cam-eras, "I'm almost convinced that the very beginning of this was by a com-munistic front."[28] Rumors also flew that young Till had been found alive in Chicago, Detroit, or New York.[29]

William Faulkner, the Nobel Prize–winning novelist and Mississippi's most celebrated son everywhere but in his home state, was of two minds, one drunk and one sober. He understood as few did the deep and global implications of the case. Asked to comment during a sojourn in Rome, he cited the "sorry and tragic error committed in my native Mississippi by two white adults on an afflicted Negro." In the perilous atomic age, amid the rise of anticolonial struggles, Faulkner said, America's Cold War competition with the Soviet Union meant that the nation could no longer afford racial atrocities and patent injustices. The Till case was the absolute nadir. "Because if we in America have reached the point in our desperate culture where we must murder children, no matter for what reason or what color, we don't deserve to survive and probably won't."[30]

Faulkner's eloquent moral resolve apparently dissolved in champagne when a reporter interviewed him in Paris a few months later. Then he declared, "The Till boy got himself into a fix and he almost got what he deserved." The interviewer asked if the murder would have been justified had Till been an adult. "It depends," Faulkner replied, "if he had been an adult and had behaved even more offensively. . . . But you don't murder a child." Soon afterward Faulkner pled drunkenness and blamed both the champagne and the interviewer. "These are statements which no sober man would make nor, it seemed to me, any sane man believe." W. E. B. Du Bois, speaking on a radio program in California, challenged Faulkner to a debate at the courthouse in Sumner. Faulkner politely declined.[31]

Elizabeth Spencer, a white Mississippi novelist a quarter century younger than Faulkner, also found herself in Rome that autumn. When she returned in September 1955 she discovered that white men had lynched Emmett Till near her father's farm in the Delta. In the expectation that her father—whose racial convictions she had regarded as enlightened—would feel much the same way, Spencer expressed her horror at the crime and the verdict. But her father "refused to discuss it," she said, "or to hear any discussion of it. He said that we [white people] had to keep things in hand." Parting ways with her father, her ears "ringing with parental abuse," Spencer headed for Oxford, Mississippi, then abandoned the state altogether and moved to New York City, resolved "in my bones, in the sick, empty feeling there inside . . . *You don't belong here anymore.*" She took with her the manuscript of her third novel, *The Voice at the Back Door*—a call for justice in Mississippi that would appear to critical acclaim in 1956.[32]

African American leaders, intent on what would happen next, underlined as Faulkner had the poisonous effect of the Till case on U.S. foreign relations in the context of the Cold War and anticolonial revolutions across the world. Here they spied an opportunity. Whites had murdered African Americans without consequence before this, and surely whites would murder African Americans without consequence after this; for them the urgent issue was how the toll could be slowed, confronted, and eventually stopped. And the international context of Emmett Till's murder offered them powerful political leverage toward that goal.

During the trial, Cora Patterson, an official in the Chicago branch of the NAACP, noted, "The eyes of the world are on that trial. It's not going to help the United States' situation in Europe and Asia if a fair trial is not held."[33] At a labor rally on Seventh Avenue in New York after the acquittals, Representative Adam Clayton Powell Jr. called the Till murder "a lynching of the Statue of Liberty. No single incident has caused as much damage to the prestige of the United States on foreign shores as what has happened in Mississippi."[34] Channing Tobias, chair of the NAACP's board, declared, "The jurors who returned this shameful verdict deserve a medal from the Kremlin for meritorious service in Communism's fight against democracy."[35]

These international dynamics were nothing new. World War II had given black Americans unprecedented power to redeem or repudiate American democracy in the eyes of the world, a fact that A. Philip Randolph employed to great effect in his wartime March on Washington movement.[36] The war also crippled European colonialism and gave rise to the Cold War rivalry between the United States and the Soviet Union. Among the darker-skinned people of the world, the postwar competition between the superpowers was urgent largely in terms of their own anticolonial concerns; therefore the caste system of color in the United States spoke louder than the nation's ringing rhetoric of democracy. A U.S. State Department report conceded that "the division of opinion on many issues" in the newly created United Nations General Assembly "has sometimes tended to follow a color line, white against non-whites, with Russia seeking to be recognized as the champion of non-whites."[37]

"It is not Russia that threatens the United States so much as Mississippi," the NAACP declared in a 1947 petition to the United Nations. The petition, which decried "the denial of human rights to minorities in the case of citizens of Negro descent in the United States," created an "international sensation," as the NAACP's Walter White put it. The national office, White said, was "flooded with requests for copies of the document" from nations "pleased to have documentary proof that the United States did not practice what it preached about freedom and democracy."[38] This new understanding that international politics might hold the key to full citizenship for African Americans marked "an historic moment in our struggle for equality," one newspaper editor wrote to W. E. B. Du Bois. "Finally we are beginning to see that America can be answerable to the family of nations for its injustices to the Negro minority."[39]

Many elements of the federal government appeared to agree with the substance of the NAACP's assertions. "We cannot escape the fact that our civil rights record has become an issue in world politics," President Truman's Committee on Civil Rights declared in 1947. "The world's press and radio are full of it. . . . Those with competing philosophies have stressed—and are shamelessly distorting—our shortcomings."[40] Secretary of State Dean Acheson wrote in 1952, "Racial discrimination in the United States remains a source of constant embarrassment to this Government in the day-to-day conduct of its foreign relations." The U.S. Justice Department filed a series of briefs in the cases leading up to the *Brown v. Board of Education* decision that supported the NAACP's position in exactly those global terms. "Racial discrimination furnishes grist for the Communist propaganda mills," Attorney General Brownell wrote to the Supreme Court, "and it raises doubts even among friendly nations as to the intensity of our devotion to the democratic faith."[41] On May 17,

1954, so-called Black Monday, the Voice of America instantly heralded the *Brown* decision across the planet in thirty-five languages.[42] Fifteen months later the Till case blew apart much of the goodwill that *Brown* had won for the United States around the globe.[43]

Newspapers in countries across the world called the acquittals "scandalous," "monstrous," "abominable," and worse.[44] Eleanor Roosevelt published an editorial entitled "I Think the Till Jury Will Have an Uneasy Conscience," in which she noted that "the colored peoples of the world, who far outnumber us," had fixed their attention on the Till trial and that the United States had "again played into the hands of the Communists and strengthened their propaganda in Africa and Asia."[45] She did not exaggerate. In fact, outrage over the verdict in Ghana represented a legitimate threat to U.S. goals in Africa.[46] Carl Rowan, a successful black journalist for *Time* magazine, was in New Delhi in the mid-1950s on a speaking tour sponsored by the State Department to highlight white Americans' growing acceptance of African Americans. He had to "face hundreds of questions over the weeks about white people in America murdering a fourteen-year-old boy named Emmett Till because he allegedly whistled at a white woman."[47] The State Department reported that "in Communist pamphleteering" in the Middle East the Till case was highlighted as "'typical' of repressive measures against minority groups in the United States, with special focus on the feature that the courts condoned this act."[48]

In 1956 the U.S. Information Agency surveyed European disdain for American race relations and found the Till case the "prevalent" concern, though it would soon be weighed alongside mob violence at the University of Alabama and in Little Rock.[49] "Since the Emmett Till case in Mississippi last fall," the American Embassy in Brussels reported to the State Department, "the Belgian Press has directed increasing attention to problems of racial relations in the United States. The press of all shades of political opinion and political leaders with whom the Embassy has talked have been astonished at and strongly condemned the racial prejudice evident in such recent cases."[50] The Italian communist press hammered readers with the Till case day after day, and the Vatican's official mouthpiece, *L'Osservatore*

Romano, found it deplorable that "a crime against an adolescent victim remains unpunished."[51] A memorandum from the American Embassy in Copenhagen to the secretary of state described "the real and continuing damage to American prestige from such tragedies as the Emmett Till case." Sweden's *Le Democrate* responded to the verdict with an editorial, "A Disgusting Parody of Justice in the State of Mississippi."[52] *Das Freie Volk* in Düsseldorf declared, "The life of a Negro in Mississippi is not worth a whistle."[53]

In 1956 the State Department reported that the "Till affairs drew greater attention in France than they did in the United States."[54] Feeling the sting of world opinion about French colonial rule in Algeria, France welcomed evidence of American hypocrisy on race. The conservative Paris daily *Figaro* printed an editorial three days after the acquittals with the banner headline "Shame on the Sumner Jury," urging Americans to "look into their own actions." The conditions under which black people live in America, observed the centrist paper *Le Monde*, should "incite more reserve and modesty on the part of those who decry the 'colonialism' of others."[55] A mass meeting in Paris adopted a resolution addressed to the U.S. ambassador calling the lynching and acquittals "an insult to the conscience of the civilized world."[56]

The *New York Post*'s editorial board perhaps said it best: "Like other great episodes in the battle for equality and justice, this trial has rocked the world, and nothing can ever be quite the same again—even in Mississippi."[57] Beyond the commonly understood borders of race, nation, and freedom, the Till case laid bare the contradictions at the heart of America's history and forced this most powerful nation to take stock of itself—if nothing else, to assess its loudly self-proclaimed standing among the world's peoples. Many believed America itself had killed Emmett Till. Its allies were concerned, its enemies gleeful. The Magnolia State's own native son, Richard Wright, issued an astute assessment of the case's implications from his self-exile in Paris: "The world will judge the judges of Mississippi."[58]

17

PROTEST POLITICS

None of the hopeful organizers of the protests on September 25, 1955, could have forecast the scope of their success. Four thousand church and United Auto Workers members packed Detroit's Bethel AME Zion, designed to accommodate 2,500; fifty thousand more lined an eight-block radius around nearby Scott Methodist Church. Representative Charles Diggs addressed between six and ten thousand people in that city, describing the "sheer perjury and fantastic twisting of the facts" at the trial in Sumner. Diggs charged that Mississippi represented "a shameful and primitive symbol of disregard for the essential dignity of all persons which must be destroyed before it destroys all that democracy is to represent."[1] Reverend C. L. Franklin, one of the most admired black preachers of his generation, showed up waving a wad of cash and urging the crowd to give generously to support the struggle. Ushers carried bags and baskets brimming with cash.[2] Churches, labor unions, and other organizations presented substantial checks. One source reported that the Diggs rally alone contributed $14,064.88 to the coffers of the NAACP, a princely sum in 1955.[3]

NAACP branches in scores of communities collaborated with more than a dozen labor unions, coordinated by the national office of the NAACP and the United Packinghouse Workers of America. The national convention of the United Electrical, Radio and Machine Workers swelled the crowd in Cleveland on September 25.[4] In Kansas City preachers and meatpackers gathered at Ward AME Church at 22nd Street and Prospect.[5]

The protest at Metropolitan Church in Chicago drew ten thousand to hear Mamie Bradley, the journalist Simeon Booker, and Willoughby Abner, president of the NAACP's Chicago branch and an official for the United Auto Workers.[6] Mamie must have quickly hopped a plane after her appearance in Chicago in order to take the stage with Roy Wilkins and A. Philip Randolph in Harlem. After gathering at churches, labor halls, and campuses across New York City, tens of thousands made their way to a big outdoor stage in Harlem to hear Randolph declare that "only the righteous revolt" of citizens throughout the country could halt "this wave of terrorism" in the South. He called for a March on Washington to protest President Dwight D. Eisenhower's failure to protect African Americans in the South. Mamie told the crowd, "What I saw at the trial was a shame before God and man."[7] An internal report of the UPWA estimated that "50,000 people turned out in New York City to hear A. Philip Randolph, president of the AFL Sleeping Car Porters, blast the whitewash of the brutal crime."[8] The Brotherhood of Sleeping Car Porters, the NAACP, and the Jewish Labor Committee, "representing 500,000 workers in the AFL," along with a number of progressive churches and other groups, sponsored the huge rally.[9]

The Mississippi underground that located nearly all the African American witnesses for the murder trial of Milam and Bryant was well represented on the stump during the September 25 national rallies. Dr. T. R. M. Howard gave a two-hour speech to a crowd of more than 2,500 at Sharp Street Methodist Church in Baltimore, calling for the investigation of negligent FBI agents in the South and irritating the hell out of Director J. Edgar Hoover. "It's getting to be a strange thing," said Howard, "that the FBI can never seem to work out who is responsible for killings of Negroes in the South."[10] In Detroit, Howard's old friend Medgar Evers also blasted the treatment of blacks in Mississippi, detailing the murders of George Lee and Lamar Smith. Ruby Hurley, who had worn a red bandana and work clothes to sneak onto the Delta plantations with Evers and Amzie Moore in search of witnesses, spoke alongside Thurgood Marshall at an "over-flow rally" in Holy Rosary Church's school auditorium in Brooklyn.[11]

A combined total of well over 100,000 attended rallies that day in Chicago, New York, Detroit, Baltimore, Cleveland, New Rochelle, Newark, Buffalo, Philadelphia, and many other cities and towns across the country.[12]

This was not the cramped, fearful McCarthy era, nor did it reflect what one Democrat called the Eisenhower administration's languid gaze "down the green fairways of indifference."[13] Something new was afoot. In cities all across America citizens found Mississippi guilty as charged. The response was a gathering of activists in the North—labor unionists, religious progressives, stalwarts of the Old Left, and ordinary citizens—that would join activists in the South to transform the Southern civil rights movement into a national coalition. The quickly emerging movement was so large that no one person or organization could manage it. Not since the Scottsboro trials of the 1930s had the country seen anything like it.[14] The killing of that boy who "did the talking" had become much more than just another lynching in the South. For some, including Mamie Bradley, it was an Archimedean lever with which to move the world.

After September 25 churches, labor unions, NAACP branches, and other organizations gathered strength for a second wave of protests the following week. The largest church in Detroit, Greater King Solomon Baptist Church at 14th and Marquette, held a twenty-four-hour service on September 29 to motivate support for rallies on October 1 and 2.[15] Twelve pastors and Mamie Bradley addressed the four thousand gathered there.

"Mrs. Bradley said she had recovered from the sorrow she felt at her son's death," reported the *Chicago Tribune*, "and 'Now I'm angry—just plain angry.' She urged a united front in the fight for civil rights."[16] The Brotherhood of Sleeping Car Porters, the Urban League, the United Steelworkers, the UPWA, and the NAACP organized mobilization meetings in Chicago, New York, Milwaukee, and Buffalo, and across the country built momentum for the massive national protests. When the day came, the headline in the *Chicago Defender* proclaimed, "100,000 Across Nation Protest Till Lynching."[17]

Mamie Bradley well knew that being the mother of Emmett Till made her uniquely useful to and powerful in the struggle. The demands on her were constant and increasingly begged trade-offs. Despite rumors that she had collapsed from exhaustion in Chicago on October 1, she walked into Williams Institutional CME Church in New York City on that same day, only a week after her last appearance there. Three thousand people had been waiting in the sanctuary, some of them for hours. The *Chicago Defender* estimated at least fifteen thousand more huddled outside, listening on loudspeakers. The throng applauded, cheered, and wept at the sight of Emmett Till's mother. A. Philip Randolph was the first to speak. "If America can send troops to Korea . . . ," he began, and the crowd drowned him out with its roar. In Cold War fashion, the New York branch of the NAACP had urged Mamie not to participate in this rally because of its alleged "pinkish tinge," or leftist sympathies, but to return instead for its own mass meeting the following Sunday. She rejected the counsel. She resented not only what the *Chicago Defender* called the "petty jealousy" and "unnecessary confusion" among the various civil rights organizers but also the huge sums being raised on her courage and her son's death while her own financial needs grew dire; she had been unable to work since being swept up by the movement.[18]

Of these concerns internal to the movement the wider world knew little and cared less. On Sunday, October 2, tens of thousands again gathered in Detroit, New York, Chicago, Baltimore, and a number of other cities to protest the Till verdict and call for federal action. Three thousand packed Chicago's Metropolitan Church and perhaps seven thousand more spilled into the street in what some said was the largest and most energetic civil rights meeting in the history of the Chicago NAACP. Willoughby Abner spoke on historic Mississippi atrocities and racial conflict in Chicago. Simeon Booker told stories of covering the trial in Sumner. Frank Brown, who had attended the trial for the UPWA, and Charles Hayes, the union's District 1 director, each held forth on the ties between civil rights and labor issues.[19] This was also the topic in Minneapolis, where the AFL demanded that the federal government act in the Till case, and at the New York convention of the International Association of Machinists, AFL, where Herbert Hill, the national NAACP labor secretary, told three hundred delegates that trade unions were imperiled "because the South remains a land of trigger-happy sheriffs and lynch mobs using violence not only against innocent Negroes but also against union organizers."[20]

The national office of the NAACP, flush with funds donated at dozens of mass meetings, scheduled rallies across the North and South in the weeks ahead, many featuring Mamie Bradley. In Alabama, it was Birmingham, Montgomery, and Tuskegee; in Georgia, Atlanta and Savannah had large protests. Both Charleston, South Carolina, and Charleston, West Virginia, were on the schedule, as were Miami and Tampa, Dallas and Fort Worth, Pittsburgh and Philadelphia, Boston and Springfield, Toledo and Cincinnati, Camden and New Brunswick. The Memphis, Nashville, Chattanooga, and Knoxville NAACP branches prepared to host rallies, as did those in St. Louis, East St. Louis, and Kansas City. Milwaukee, Des Moines, and Washington, D.C., rounded out the tour. Whether Mamie would bear up under the pace of this schedule remained to be seen.[21]

For the rest of October and in November protests continued from New York to Los Angeles. Moses Wright spoke from time to time, walking back and forth, pounding his fist into his palm, and animating his performances in the manner of a veteran Church of God in Christ preacher. "Rally of 20,000 here cheers call for action against Mississippi goods," the *New York Times* reported on October 12. Jammed into the Garment District on 36th Street between Seventh and Eighth, the twenty thousand protestors roared their approval when Adam Clayton Powell Jr. proposed a national boycott on Mississippi products and a March on Washington in January to demand that Congress finally pass an antilynching bill. Activists quickly organized March on Washington Committees in New York, Detroit, Chicago, and elsewhere.[22]

Throughout the fall and into the winter, mass meetings flourished in New York, Chicago, and Detroit, as well as Newark, Boston, Cleveland, St. Louis, and Milwaukee.[23] In Los Angeles some five thousand persons crowded into the Second Baptist Church, with several thousand more standing outside, to protest the Till case. Almost certainly a majority of them were first- or second-generation Southern transplants of the Great Migration. As described by the *Chicago Defender*, "The meeting was punctuated by screams and occasional outcries as Dr. T. R. M. Howard of Mississippi related facts and details about what he termed the 'cruel, desperate and deadly' treatment of Negroes in his home state." Silence ruled when the eloquent physician turned his eye to Los Angeles itself. "How many of you within my sight and the sound of my voice are enjoying the luxury of Cadillacs, safe, comfortable homes, and the privilege to vote, while thousands of your brothers in blood live in fear of their lives?" he demanded. The rally raised roughly $10,000.[24]

In mid-November the NAACP disclosed that more than a quarter million people had heard Mamie Bradley or Moses Wright speak at their rallies, and Howard claimed that he had addressed thirty thousand people.[25] When Medgar Evers reported on his out-of-state speeches on the Till case, he listed Detroit twice, St. Louis, East St. Louis, Washington, D.C., Tampa, and Nashville.[26] Both Howard and Adam Clayton Powell spoke in Montgomery in November. There would be many more speeches in a movement organized by indignation over the Till case.

Any thought that the case would fade as 1955 turned into 1956 would soon vanish, thanks to the ongoing protests but also to the work of William Bradford Huie, a seventh-generation Alabama novelist and journalist with an inflamed ambition and iridescent imagination. In January 1956

Look magazine, with one of the largest circulations in the country, published Huie's "Shocking Story of Approved Killing in Mississippi."[27] In addition to copies for their nearly four million subscribers, *Look* printed an extra two million for the newsstands.[28] Three months later the story was reprinted for eleven million subscribers to the *Reader's Digest*.[29] Huie's story would shape America's imagination of the Till case for fifty years.

Huie began working on the Till case a month or so after the acquittals.[30] In Sumner he met with J. J. Breland, pointing out that the truth of what happened had never been established. "And this lawyer said, 'Well, I'd like to know what happened. I never asked them whether they killed the boy or not.' "[31] A pioneer in what would later be derided as "checkbook journalism," Huie told Breland that *Look* would pay Milam and Bryant $4,000 to give their account. Breland called in the killers and relayed Huie's offer. Because they had been acquitted, they could not be tried again for the same crime; therefore, without shame or law to impede them, and with cash on the table, there was no reason not to go public with their version of events. They accepted. There would be $1,000 for the law firm and $3,000 to be divided between Milam and Bryant in exchange for the story of Till's kidnapping, beating, and murder. Huie would state the facts, including quotes, without saying how he had gotten them; that would allow the half-brothers to maintain some pretense of innocence. And they would sign a waiver not to sue Huie for libel. Breland then set up a week of secret meetings at the law firm at night. After *Look*'s senior counsel showed up with a satchel full of cash, J. W. Milam and Roy and Carolyn Bryant told Huie their story through a haze of cigarette smoke, with Milam doing most if not all of the talking.[32]

If any of them mentioned a physical assault of any kind on Carolyn, Huie did not report it, which seems unlikely given his penchant for the sensational. In this version Emmett Till of Chicago, visiting his country kinfolks in Mississippi, boasted to his young cousins about having had sex with a white girl. Outside Bryant's Grocery the youths dared Till to ask Carolyn Bryant for a date. He did so. Hearing the tale, Milam and Bryant kidnapped the boy from his great-uncle's farmhouse, intending merely to beat him, but Till taunted them with stories of having sex with white girls and proclaimed his own equality. In short, the boy virtually committed suicide.

"We were never able to scare him," Milam told Huie. "They had just filled him so full of that poison he was hopeless." The men took turns smashing Till across the head with their .45s. The boy never yelled, but continued to say things like "You bastards. I'm not afraid of you. I'm as good as you are. I've 'had' white women. My grandmother was a white woman." Milam made their case:

> Well, what else could we do? He was hopeless. I'm no bully; I never hurt a nigger. I like niggers—in their place—I know how to work 'em. But I just decided it was time a few people got put on notice. As long as I live and can do anything about it, niggers are gonna stay in their place. Niggers ain't gonna vote where I live. If they did, they'd control the government. They ain't gonna go to school with my kids. And when a nigger even gets close to mentioning sex with a white woman, he's tired o' livin'. I'm likely to kill him. Me and my folks fought for this country and we have some rights. . . . Goddam you, I'm going to make an example of you—just so everybody can know where me and my folks stand.[33]

And so, Milam said, they drove to the cotton gin, forced Till to carry the heavy fan to the truck, took him to the riverbank, shot him in the head, and rolled him into the river.[34] In this version, it was a "coincidence" that Till's ignorant bravado met Milam's ignorant brutality at just the wrong place and time, "too soon after the Supreme Court had decreed a change in the Delta 'way of life.'"[35] This was not a tell-all; for one thing, Huie knew that more than two people were involved in the murder of Emmett Till, but he decided to forget that inconvenient fact because it would cost too much to get releases to print their names.[36] In Milam and Bryant's version the one person who receded further from view was the real-life fourteen-year-old Emmett Till, with his slight stutter, his imitations of Red Skelton and Jack Benny, and his ability to see second base in a loaf of bread.

• • •

The real Emmett Till was also not in evidence at the national protests that continued well into the spring of 1956, but his family and their Mississippi allies certainly were. T. R. M. Howard's finest hour came at what A. Philip Randolph billed as "the Historic Madison Square Garden Civil Rights Rally," otherwise known as the "Heroes of the South" rally. The Brotherhood of Sleeping Car Porters did much of the organizing, and In Friendship, a fundraising organization, did the rest. Ella Baker, Bayard Rustin, Stanley Levison, Norman Thomas, and other liberals, radicals, and labor activists in New York founded In Friendship in early 1956, using the energy around the Till case and the Montgomery Bus Boycott to fund the boycott and to assist grassroots activists in the South who suffered economic reprisals for their civil rights activities. The Heroes of the South rally, set for May 24, 1956, at Madison Square Garden, was their first big project.[37]

The rally drew more than sixteen thousand people. Besides Randolph and Howard, the evening featured Eleanor Roosevelt, Adam Clayton Powell Jr., Roy Wilkins, Autherine Lucy, Rosa Parks, and E. D. Nixon. Parks and Nixon filled in for the headliner, Martin Luther King Jr., who backed out to attend a meeting in Montgomery. Sammy Davis Jr. and the bandleader Cab Calloway provided musical entertainment.

Randolph introduced Howard as "a man who has dedicated his life to the cause." He praised Howard for confronting "the racialism and tribalism of those who would strike down the Constitution and take away the rights of the people of Mississippi merely because of color."[38] Howard followed with a Cold War thrust: "I come to you tonight from that Iron Curtain country called Mississippi. I tried to call home a few minutes ago to see if we were still in the Union." He quickly reviewed his home state's hardcore racist history, mentioning the notorious Senator Theodore Bilbo and the still venomous Senator James O. Eastland. His humor was sly and sharp-edged. "The governor recently called a special session of the legislature to have the letters N, A, C and P removed from the alphabet and to have the words 'integration' and 'desegregation' stricken from the English language."[39]

Howard described the recent upheavals in race relations in Mississippi since *Brown v. Board of Education* and the NAACP's campaigns for desegregation of the public schools and access to the ballot. The Citizens' Council, "the worst internal threat that we have to our American way of life," was spreading across the South. Reverend George Lee, "a great friend of mine," and Lamar Smith, "a personal friend of mine," had been murdered for seeking voting rights, and no one had been arrested in either case. The boy Emmett Till had ascended to a pantheon of martyrs, a rallying point alongside Lee and Smith for justice and progress. "We have grown tired of fighting for something across the sea that we can't vote for in Belzoni." Howard asserted, "The [white] people of Mississippi are not afraid of the Supreme Court. They are not afraid of Congress. But every time you mention the NAACP, they tremble in their boots. We should all support the National Association for the Advancement of Colored People."

Asking the crowd for a show of hands, Howard remarked that a large majority of those present were blacks born in the South. "The reason that you're in New York," he said, "is because you were running away from some of those conditions that we have described here tonight." He confronted not just the Jim Crow South but the complacent North: "The pitiful part of it is that you have forgotten about the conditions in the South and you are not doing much about the damnable conditions that exist right here in New York City." Appreciating his candor, the crowd slowly began to applaud, building to a roar.

"Ladies and gentlemen," he responded, "the people of South Carolina, Georgia, Alabama, and Mississippi cannot hear your applause tonight. We've got to do something more than just clap our hands about the situation. I believe the papers ought to carry tomorrow that this group gave a hundred thousand dollars to help carry the fight in the Southland."[40]

As Howard took his seat, Randolph reminded the crowd that they were "planning to use the funds that are collected at this meeting for the National Association for the Advancement of Colored People and the Montgomery Bus Boycott." At that point, the boycott was almost six months old and showed no sign of ending anytime soon. The UPWA donated $1,632.04, boosting their support for the boycott that spring to well over $5,000; they would continue to be major supporters of Dr. King's work throughout the 1960s. Local 32 of the United Steelworkers gave an additional $1,000 at the rally, and the Transport Workers Union wrote a check for $700.[41]

Conspicuous by her absence that night was Mamie Bradley. She had shared her grief with the nation, inviting the world to "see what they did to my boy." In early September, weeks before the beginning of the trial, she had clarity about her role in the drama of her boy's death and the potential redemption of her loss. "This is not just for Emmett," she said only a week after his funeral, "because my boy can't be helped now, but to make it safe for other boys." She vowed to see that his killers were punished but also insisted that the federal government must protect all black citizens. "I'm willing to go anywhere, to speak anywhere, to get justice." A falling-out with Roy Wilkins soon curtailed her speaking schedule, but she never ended her cry for justice, not until she died in 2003.[42]

Except there was no justice to be had for Emmett Till, not in 1955, not in 1956, not even decades later. The depth and malignancy of white supremacy beggared notions of justice. There was progress, eventually, which came with the effort of a great many and the deaths of more than a few on the battlefields of race in America. One consequence of the public battle that Mamie Bradley started was that her son became less of a boy and more of an icon that spoke to those struggling across the nation, in both the North and the South. "Everyone knew we were under attack," recalled one black journalist, "and that attack was symbolized by the attack on a fourteen-year-old boy."[43]

White Southerners regarded this obsession with Emmett Till as an injustice to them. They pointed to the hypocrisy of white Northerners who protested the Till murder but refused to see the racial brutality around them. When Mamie and Mayor Richard Daley of Chicago joined in asking President Eisenhower to intervene in the Till case, the contradictions were too much for some Mississippians. Their statement, reported the *Greenwood Morning Star*, was something that "Mississippi people resent

very much. We notice that a press dispatch says there have been 27 bombings in Chicago in the past 16 months and that they have gone unsolved."
The mayor of Chicago should "clean up his own crimes before he spouts
off in condemnation of Mississippi."[44]

That was a sentiment more or less shared by many African Americans
in Chicago. The Packinghouse Workers endorsed Mayor Daley's appeal
to Eisenhower but pointed out that "Emmett Till was hardly less safe in
Greenwood, Mississippi than he would have been in Trumbull Park, Chicago," where racial clashes over housing continued.[45] If some in the North,
Daley first among them, could be accused of protesting white supremacy
in the South and protecting it in the North, many in Chicago were coming to understand full well that the battles fought in the North and in the
South were all part of the same war. "We have our own Mississippi only
20 minutes away from here," declared Abner Willoughby, one of the main
organizers of protests at Trumbull Park.[46]

If the protest politics around the Till case brought neither justice nor
clarity, they brought together many elements that would soon anchor the
civil rights movement. What happened to Emmett Till in Mississippi had
helped galvanize what was now a national movement. Whatever their
other differences, the Till case joined activists north and south in a common struggle, giving those in Chicago, Detroit, New York, and other
Northern cities a stake in the racial politics of a distant region. It united
and pulled into the movement black labor unions and integrated unions
like the UPWA that were rapidly becoming "civil rights unions," whether
or not all of the white members agreed with integration. The fragmented
American left made common cause in the Till case. Many religious organizations became players in the effort to support the movement. The bitter
realities of Emmett Till's lynching and the acquittals gave civil rights a
new moral appeal that it bears to this day. The NAACP's labor director
Herbert Hill mused years later that even if the Till case had only helped
form "a Northern consensus about Southern racism," that consensus was
invaluable to the growing civil rights movement in Dixie and across the
country.[47]

18

KILLING
EMMETT TILL

If the past is irrevocably gone, and we cannot somehow conjure it back and see it with the omniscient eyes of God, we can nevertheless follow wherever the fragments of evidence lead us and try to understand what they tell us. What we know happened to Emmett Till is cause aplenty for sorrow and anguish; the mysteries that persist mean little or nothing for the insistent dilemmas of race in America. Exactly what happened between Carolyn Bryant and Emmett Till at the store will never be revealed to a certainty, probably not even to her. What she did or didn't do with respect to his kidnapping may never be entirely clear. But how Mamie Bradley dug deep within herself and inspired thousands of other Americans to move is clear enough. From this tragedy large, diverse numbers of people organized a movement that grew to transform a nation, not sufficiently but certainly meaningfully. What matters most is what we have done and will do with what we do know. We must look at the facts squarely, not to flounder in a bitter nostalgia of pain but to redeem a democratic promise rooted in the living ingredients of our own history. The bloody and unjust arc of our history will not bend upward if we merely pretend that history did not happen here. We cannot transcend our past without confronting it. As Du Bois wrote in 1912, "This country has had its appetite for facts on the Negro question spoiled by sweets."[1]

So, the facts: Emmett Till's executioners' ghastly errand began at an undisclosed location that could have been as far away as J. W. Milam's store in Glendora, thirty miles distant. J.W., Roy Bryant, their brother-in-law Melvin Campbell, and a friend, probably Hubert Clark, were playing cards and drinking hard when the subject of what J.W. called the "smart talk" and whistling incident between Carolyn and "the Chicago boy" came up. Two black men who worked for J.W., Henry Lee Loggins and Levi "Too Tight" Collins, were likely present, though not part of the card game. The white men agreed that this affront at Bryant's Grocery could not go unavenged. They decided to go get the boy. According to some accounts, J.W. borrowed Clark's old car because his brand-new two-toned Chevrolet pickup would be too recognizable.[2] It is possible that Carolyn went to the Wright house with them and identified Emmett before they kidnapped him, so they may have stopped at Bryant's Grocery to pick up Carolyn on the way. Or she may have remained at the store and either identified him there or refused to do so when they came back with Emmett. Although her accounts of this night have been inconsistent over the years, she has always maintained that she told her husband the boy they brought to the store for her to identify was not Emmett Till. Whether or not she actually identified the boy is merely a matter of speculation; I have found no way to prove or disprove it. The preponderance of evidence does tell us that almost from the moment of the incident between her and Emmett at the store on August 24, she was frightened of its escalating consequences and probably sought to avoid them.

After they kidnapped Emmett, and whatever may have happened at Bryant's Grocery that night, the white men hauled the black teenager with them to the place where they had been drinking. Present were J.W., Roy,

Campbell, and Clark or perhaps Elmer Kimbell. Some combination of black men who worked for J.W.—possibly Loggins, Collins, Otha "Oso" Johnson, and Joe Willie Hubbard, perhaps only one or two of them—rode in the back of the truck with Emmett. Back at their clandestine watering hole, the white men beat and berated Emmett for as much as an hour. He was neither dead nor unconscious when they decided to take the boy to a hundred-foot bluff overlooking the Mississippi River near Rosedale, a place J.W. knew about, perhaps to scare him, though it is just as likely they intended to kill him there and dump him in the river. The men piled into J.W.'s truck for the hour-long drive to Rosedale. The four white men rode in the large cab. One or more of the black men stayed in the bed of the truck to keep Emmett from fleeing.

By some accounts Collins was alone with Emmett in the back and had a hard time controlling the boy, so they stopped at a juke joint in Glendora and picked up some of the other men who worked for J.W.: Hubbard, Loggins, and perhaps Johnson. At this distance in time the role of the black men is a little hard to fathom. They could have had few illusions about the fate of the boy they were restraining. Their behavior may reflect their terror of and utter subservience to J.W.; they would have known that their objections to the boy's fate would carry no weight and that the white men could kill them with impunity at any point. There was room in the Mississippi or the Tallahatchie for their bodies, too. An African American's testimony in a court of law was all but useless in 1950s Mississippi, and they were unlikely to report the crime to Sheriff Strider. It is also possible that these black men suffered from an internalized white supremacy so deep that they virtually never questioned the prerogatives of white men; their world was certainly constructed to make it so. Suffice it to say that at every juncture the white men were calling the shots.

White men in the front, black men in the back, they drove around for some time, looking for J.W.'s proposed spot on the Mississippi, but failed to locate it for an hour or more. The drunken white men then turned the truck toward the farm that Leslie Milam managed, "tryin' to make our minds up," Roy told an interviewer, about what to do with the boy. There was a large equipment shed at the place where they could continue to torture him and perhaps decide his fate. It may not have been a foregone conclusion, at least not among all of them, that they would kill him when they got there. The truck rumbled onto the farm soon after sunrise. Leslie Milam was not happy to see them, in part because he had work to do that day, but he agreed to let them use the shed and then joined them there.[3]

Once inside the large equipment shed the men grabbed Emmett off the truck and began to beat him again, this time with shattering force. Most of the blows were to the boy's head, their principal weapons the large-caliber pistols they carried. J.W. wore on his belt a heavy Ithaca-brand, U.S. Army–issued .45 semi-automatic, a hunk of steel that weighed 2.7 pounds, heavier than most carpenter's hammers, and Roy carried a similar pistol. It is likely that one or more of the men used tools found in the shed; observers described wounds to the left side of Emmett's face that seemed to have been made with a heavy blade.

While no one outside the shed saw what happened inside, several witnesses heard it. Willie Reed was awakened by his grandfather Add Reed at dawn and sent walking to the store. Passing near the barn, Reed first saw the four white men in the cab of the truck and three black men and a boy in the back. Cutting across the field, he heard the brutal sounds of a beating and agonized cries for mercy. As he came closer to the barn, a big bald-headed man emerged and walked to the nearby well for a drink of water. Reed later identified the man as J. W. Milam. "He had on a pistol. He had it on his belt." J.W. stepped quickly back into the shed. According to T. R. M. Howard, Reed told him that he heard the frightened boy crying from the barn, "Mama, please save me," and "Please, God, don't do it again." Reed testified in court, "I heard somebody hollering and I heard some licks like somebody was whipping somebody." As to the number of blows: "There was a whole lot of them." The intense screams finally faded to whimpers, then stopped altogether. Add Reed and Mandy Bradley, a neighbor who lived close by, supported Willie Reed's account of events.[4]

The merciless ferocity of the assault may be proven by the injuries to

Emmett Till's body. Initially Emmett tried to fend off the blows with his hands and arms, but soon he was unable to do so. Both of his wrists were broken in the effort to defend himself. Vicious blows crushed in the whole crown of his head; similar strokes smashed the back of his skull so that pieces of it would fall off as the law enforcement officers pulled him from the river. Surprisingly, given Mamie Bradley's description of his smashed teeth, only one of his teeth was missing; nor was he castrated, as some other accounts allege. His assailants managed to fracture his thigh bone, the largest and strongest bone in the body, which suggests they stomped on him with enormous and repeated force or beat him with something much heavier than a pistol. Deputy Sheriff John Ed Cothran testified that "his left eye was about out, it was all gouged out in there." Part of one ear was missing, which could have meant someone tortured him with a knife or shears. A farm equipment shed likely offered a number of choices for implements to inflict pain and cause death. The injuries to his head almost certainly would have killed him, but the immediate cause of death was a gunshot wound above his right ear.[5]

Amid all informed speculation, there is this fact: it takes from five hundred to one thousand pounds of force to crack a human skull. No one exerts that level of force on a fourteen-year-old's head without willingness to kill. When Carolyn Bryant says Emmett Till didn't deserve what happened to him, this—delivering hundreds of pounds of force at impact, over and over again—is part of "what happened to him," in addition to the other unspeakable tortures that went on in that barn.[6] Clipped ear. Broken bones. Gouged-out eye. The ruthless attack inflicted injuries almost certain to be fatal. They reveal a breathtaking level of savagery, a brutality that cannot be explained without considering rabid homicidal intent or a rage utterly beyond control. Affronted white supremacy drove every blow.

Despite the viciousness of the assault, it is not clear exactly when the men consciously made up their minds to kill the boy. In Bryant and Milam family lore, true or not, the story goes that at some point Roy developed misgivings about the fatal beating. According to Carolyn, the men told her that Roy had wanted to stop and take Emmett's broken body to the hospital. She said this would have meant dumping his body in front of a medical establishment and driving away. "Well, we done whopped the son of a bitch," Roy told a friend in 1985 who was wearing a hidden recording device, "and I had backed out on killing the motherfucker." In the end, Roy told his friend, they decided that "carryin' him to the hospital wouldn't have done him no good" and instead they would "put his ass in the Tallahatchie River." The proposal to take Emmett to the hospital, Carolyn told me, violated the sensibilities of Melvin Campbell, who muttered a curse and fired a .45-caliber bullet into Emmett's brain. This may have been only a final malignant gesture, given the boy's injuries. But certainly the gunshot brought an emphatic end to the grisly proceedings.[7]

J.W. ordered the black men to clean the floor of the shed thoroughly of blood and to spread cotton seed to cover it. After stripping the body, J.W., Roy, and Campbell put it in the truck bed and covered it with a tarpaulin. Either Clark or Kimbell borrowed Leslie Milam's car and took the black men to bury the bloody clothing. J.W., Roy, and Campbell, perhaps taking barbed wire from the shed and a heavy gin fan from a nearby building, drove ten miles across the county line to the Tallahatchie River, lashed the fan around his neck with the wire, and rolled the body into deep water.[8]

A young black man and his father reported walking past J.W.'s store in Glendora that Sunday morning and seeing J.W.'s new truck parked next to the building. A tarpaulin covered the bed, but blood had pooled on the ground. Johnson and Collins stood guard, one with his foot on the tarp and the other standing by the truck. The young man recalled that when J.W. came out of the store and his father commented on the blood, J.W. replied that he had killed a deer. When the father pointed out that it was not deer season, J.W. pulled him over to the truck, yanked back the tarpaulin, and said, "This is what happens to smart niggers." Speechless, the father turned, grabbed his son by the shoulder, pulled him toward home, and never told his son what he had seen, whether an enormous amount of blood or the corpse itself. One of the main routes from Leslie Milam's farm to the Tallahatchie River runs through Glendora, and the men might have stopped at the store on their way to dump Emmett's body.[9]

These are the facts. But we are obliged to go beyond the facts of the lynching and grapple with its meaning. If we refuse to look beneath the surface, we can simply blame some Southern white peckerwoods and a bottle of corn whiskey. We can lay the responsibility for Emmett Till's terrible fate on the redneck monsters of the South and congratulate ourselves for not being one of them. We can also place, and over the decades many of us have placed, some percentage of the blame on Emmett, who should have known better, should have watched himself, policed his thoughts and deeds, gone more quietly through the Delta that summer. Had he only done so, he would have found his way back to Chicago unharmed. That we blame the murderous pack is not the problem; even the idea that we can blame the black boy is not so much the problem, though it carries with it several absurdities. The problem is *why* we blame them. We blame them to avoid seeing that the lynching of Emmett Till was caused by the nature and history of America itself and by a social system that has changed over the decades, but not as much as we pretend.

In "Letter from a Birmingham Jail," Martin Luther King Jr. writes that his worst enemies are not the members of Citizens' Councils or the Ku Klux Klan but "the white moderate" who claims to support the goals of the movement but deplores its methods of protest and deprecates its timetable for change: "We will have to repent in this generation not merely for the vitriolic words and actions of the bad people but for the appalling silence of the good people."[10]

When we blame those who brought about the brutal murder of Emmett Till, we have to count President Eisenhower, who did not consider the national honor at stake when white Southerners prevented African Americans from voting; who would not enforce the edicts of the highest court in the land, telling Chief Justice Earl Warren, "All [opponents of desegregation] are concerned about is to see that their sweet little girls are not required to sit in schools alongside some big, overgrown Negroes."[11] We must count Attorney General Herbert Brownell Jr., who demurred that the federal government had no jurisdiction in the political assassinations of George Lee and Lamar Smith that summer, thus not only preventing

African Americans from voting but also enabling Milam and Bryant to feel confident that they could murder a fourteen-year-old boy with impunity. Brownell, a creature of politics, likewise refused to intervene in the Till case. We must count the politicians who ran for office in Mississippi thumping the podium for segregation and whipping crowds into a frenzy about the terrifying prospects of school desegregation and black voting. This goes double for the Citizens' Councils, which deliberately created an environment in which they knew white terrorism was inevitable. We must count the jurors and the editors who provided cover for Milam, Bryant, and the rest. Above all, we have to count the millions of citizens of all colors and in all regions who knew about the rampant racial injustice in America and did nothing to end it. The black novelist Chester Himes wrote a letter to the editor of the *New York Post* the day he heard the news of Milam's and Bryant's acquittals: "The real horror comes when your dead brain must face the fact that we as a nation don't want it to stop. If we wanted to, we would."[12]

Emmett Till's death was an extreme example of the logic of America's national racial caste system. To look beneath the surface of these facts is to ask ourselves what our relationship is today to the legacies of that caste system—legacies that still end the lives of young African Americans for no reason other than the color of their American skin and the content of our national character. Recall that Faulkner, asked to comment on the Till case when he was sober, responded, "If we in America have reached the point in our desperate culture where we must murder children, no matter for what reason or what color, we don't deserve to survive and probably won't."[13] Ask yourself whether America's predicament is really so different now.

EPILOGUE

THE CHILDREN OF EMMETT TILL

Emmett Till is dead. I don't know
why he can't just stay dead.
—ROY BRYANT, quoted in Mamie Till-Mobley and
Christopher Benson, *Death of Innocence: The Story
of the Hate Crime That Changed America*

The struggle of humanity against power is always
the struggle of memory against forgetting.
—MILAN KUNDERA, *The Book of Laughter and Forgetting*

"I am sure you read of the lynch-murder of young Emmett Till of Chicago," Rosa Parks wrote to a friend soon after these notorious events. "This case could be multiplied many times in the South, not only Miss., but Ala, Georgia, Fla."[1] A month after the acquittals Parks joined an overflow crowd at Martin Luther King's Dexter Avenue Baptist Church in Montgomery to hear Dr. T. R. M. Howard speak on the Till case. Dr.

King introduced the fiery physician from Mound Bayou, who spoke passionately about the assassinations of George Lee and Lamar Smith and told the story of Emmett Till in riveting detail.[2] Parks had read about the lynching and wept at the gruesome photograph in *Jet*, but Howard's story was a powerful firsthand account. His speech moved her deeply and preoccupied her for days. Only four days later Parks defied the segregation laws on a Montgomery city bus. The driver insisted that she move. Thinking of Emmett Till, she said, Parks refused to do so. Her subsequent arrest provided the occasion for the Montgomery Bus Boycott.[3]

The impact of the Till lynching resonated across America for years, touching virtually everyone who heard it, but the case had its most profound effect on a generation of African Americans two decades younger than Parks. Folks of all ages discussed the Till case in barbershops, churches, and living rooms north and south. For black youth across the country, however, the Till lynching became a decisive moment in the development of their consciousness around race. "The murder just shocked me," recalled Kareem Abdul-Jabbar, a legendary star in the NBA and later an accomplished author. "I began thinking of myself as a black person for the first time, not just a person." Muhammad Ali recalled the effect of Till's death on him: "I realized that this could just as easily have been a story about me or my brother." Gary's Richard Hatcher, the first African American elected mayor of a major U.S. city, said the killing made him "very bitter and angry towards white people. That's how I felt at that particular time."[4] A woman who grew up in Chicago and Hattiesburg, Mississippi, and later became a civil rights activist remembered, "The first time I was really confronted with this black-white issue was with Emmett Till. That really slapped me in the face."[5]

Joyce Ladner, a Mississippi native who became a renowned activist with the Student Nonviolent Coordinating Committee (SNCC), called herself and the other young blacks who grew up in the 1950s "the Emmett Till generation."[6] Charles McDew, eventually chair of SNCC, said that all the young people in the movement "knew where they were when they saw the pictures of Emmett Till's body."[7] For Julian Bond, who worked with SNCC and later became chair of the national board of the NAACP, the Till case was one of the key events that "provided stepping stones leading inexorably toward my involvement in the freedom struggle." He was "moved along the path to later activism by the graphic pictures that appeared in *Jet* magazine of Emmett Till's swollen and misshapen body."[8] Fay Bellamy Powell, who later worked with SNCC, was horrified at the coverage of the Till case and yet steeled for the struggles to come: "My spirit allowed me a glimpse of the future, saying, 'Don't worry about this. You will have an opportunity to address this madness. You will assist in showing the world the face of this evil.'"[9]

With style and daring, "the Emmett Till generation" showed the world a great deal when they launched the sit-ins that swept across the South in the spring of 1960. Four freshmen at North Carolina A&T walked into the Greensboro Woolworth's department store on Monday, February 1, bought a number of personal items, then sat at the segregated lunch counter and asked to be served. The four remained for almost an hour, until the lunch counter closed. On Tuesday twenty-five men and four women, all students at A&T, occupied the lunch counter. On Wednesday sixty-three students participated, joined in the afternoon by three white students from Greensboro College. On Thursday hundreds more black students became involved. By the end of the week students in other cities and towns in North Carolina—Raleigh, Charlotte, Fayetteville, High Point, Concord, and Elizabeth City—sat in at segregated lunch counters. By the end of the month young people across the South were organizing sit-ins. Within two months the demonstrations had spread to fifty-four cities in nine states; within a year more than a hundred cities witnessed similar protests. A new, mass-based phase of the civil rights movement, a distinctive radicalism rooted in nonviolent direct action, had begun.[10] Driving it were young people, many of whom had been inspired to action by the story of a boy their age lynched in Mississippi.

Six decades later a white police officer shot and killed a young black man named Michael Brown in Ferguson, Missouri. The local grand jury's decision not to prosecute the officer expanded and enraged a national movement born of similar killings, and young protestors throughout the United States chanted, "Say his name! Emmett Till! Say his name! Emmett Till!" His name, invoked alongside a litany of the names of unarmed black men and women who died at the hands of police officers, remained a symbol of the destructiveness of white supremacy. Hundreds of young people thronged the fence in front of the White House, chanting, "How many black kids will you kill? Michael Brown, Emmett Till!" Black Lives Matter—a movement, not just a hashtag—quickly became symbolic shorthand for the struggle. Much like the Emmett Till protests of the 1950s, these demonstrations raged from coast to coast and fueled scores of local campaigns. Police brutality against men and women of color provided the most urgent grievance but represented a range of festering racial problems: the criminalization of black bodies; the militarization of law enforcement; mass incarceration; racial injustice in the judicial system; the chasms of inequality between black and white and rich and poor; racial disparities in virtually every measure of well-being, from employment and education to health care. Such moments could make movements, wrote the political scientist Frederick Harris in the *Washington Post*, and become "like the murder of 14-year-old Emmett Till in 1955 . . . transformative episodes that remake perceptions and force a society to abandon abhorrent practices."[11]

On November 17, 2014, as these protests spread across the country, a former chair of SNCC stood in the rain on the grounds of the U.S. Capitol with a shovel. In an orchard of umbrellas Representative John Lewis helped plant an American sycamore in honor of a fourteen-year-old boy from Chicago who was murdered almost sixty years earlier. In his 1998 memoir, Lewis writes that when he was fifteen, "and at the edge of my own manhood just like him," he had been "shaken to the core" by the lynching of Emmett Till.[12] Among those wielding shovels with Lewis were both U.S. senators from Mississippi and Eric Holder, the first African American U.S. attorney general.

"Even today, the pain from this unspeakable crime, this unspeakable tragedy, still feels raw," Holder declared, but the tree would become

Emmett Till's "living memorial, here at the heart of our Republic, in the shadow of the United States Capitol." Till perished senselessly and far too soon, the attorney general said, but "it can never be said that he died in vain. His tragic murder galvanized millions to action."[13] After Holder spoke, reporters asked him about the relationship between Emmett Till and the contemporary racial conflagrations in Ferguson and elsewhere. "The struggle goes on," replied Holder. "There is an enduring legacy that Emmett Till has left us with that we still have to confront as a nation."[14]

Decades after his death Emmett continues to be a national metaphor for our racial nightmares. And difficult though it is to bear, his story can leave us reaching for our better angels and moving toward higher ground. By suffering comes wisdom, the ancient Greeks tell us, and Mamie Bradley's decision to take history in her hands and help build a movement distills that harder wisdom and leaves it in our own. "The struggle of humanity against power," writes Milan Kundera, "is always the struggle of memory against forgetting."[15]

America is still killing Emmett Till, and often for the same reasons that drove the violent segregationists of the 1950s and 1960s. Yes, many things have changed; the kind of violence that snatched Till's life strikes only rarely. A white supremacist gunman slaughtering nine black churchgoers in a prayer meeting in Charleston, South Carolina, in 2014, however, reminds us that the ideology of white supremacy remains with us in its most brutal and overt forms. "You rape our women and you're taking over our country," the murderer said as he fired round after round into his African American victims. He could have been quoting Judge Thomas Brady's 1954 *Black Monday* or a Reconstruction-era political pamphlet. White America's heritage of imagining blacks as fierce criminals, intent on political and sexual domination, as threatening bodies to be monitored and controlled, has never disappeared. These delusions have played a compelling and bloody role for centuries. The historian Stephen Kantrowitz writes that the murders in Charleston are "an expression and a consequence of American history—a history that the nation has hardly reckoned with, much less overcome."[16]

Evidence that the past is still with us is abundant. Racial hatred drove a group of suburban white teenagers to beat and murder James Craig Anderson, a black man selected at random in Jackson, Mississippi, on June 26, 2011. One teenager yelled "White power" as he returned from the assault, and several of the others shouted racial epithets. Federal and state courts charged ten of the youths with the murder and with a series of similar attacks over a period of months. The judges sentenced one of them to two life terms and the others to terms ranging from eighteen and a half months to four years.[17] Denying that such violence is common, the Hinds County district attorney said in 2011, "I do think because of the political and economic structure and the re-engineering of society, it appears that certain parts of the country and Mississippi feel their culture is under attack."[18]

Certainly politics in the United States in the first two decades of the twenty-first century reflect, as the nominal gains of the civil rights movement continue to make their claims on our society—most notably the election of America's first African American president—that many white citizens feel that something has been and is being taken from them. Nearly forty years of stagnating wages and growing inequality have done nothing to ease their anxieties. Many, too, are afraid and seek to build walls rather than bridges between our increasingly divided nations, separate, unequal, and often hostile; immigration has rendered our predicaments increasingly complex—and, for many, more frightening. Most African American children grow up in a world far more impoverished, bleak, and confined than their white counterparts. Their families fall behind white families in virtually every measure of well-being: wealth and income levels, wages, unemployment rates, health and mortality figures, levels of incarceration, and crime victimization rates. And their often lethally divergent experiences with law enforcement only mirror what Maya Angelou calls "These Yet to Be United States."[19]

America is still killing Emmett Till, but often by means less direct than bludgeons and bullets. The most successful killers of African American youth are poverty, resegregated and neglected public schools, gang violence, and lack of economic opportunities. Violence and exploitation against black women scar whole communities, and mothers still bear the burden of burying black sons. In many inner cities the drug trade is the only enterprise that is hiring, while the national unemployment rate for young black men is well over twice that for other young men. The so-called war on drugs successfully targets young African American men, even though blacks and whites use and sell illicit drugs at roughly the same rate. The enormous incarcerated and judicially supervised population of the United States has become disproportionately a population of color. Writes Ta-Nehisi Coates, "A society that protects some people through a system of schools, government-backed home loans, and ancestral wealth but can only protect you with the club of criminal justice has either failed at enforcing its intentions or has succeeded at something much darker."[20]

African American males experience the highest imprisonment rate of all demographic groups. In Washington, D.C., the country's capital, roughly 75 percent of young black men can anticipate serving time in prison, and the percentage is still higher in the city's poorest neighbor-hoods. The criminal justice system in some states imprisons black men on drug charges at rates twenty to fifty times higher than those of white men. In major cities where the drug war rages, as many as 80 percent of young black men have criminal records and thus can be legally discriminated against in housing, employment, and often voting for the rest of their lives. These statistics reflect the emergence of a new racial caste system, born from the one that killed Emmett Till.[21] "While the blame for the grisly mutilation of Till has been placed upon two cruel men," Martin Luther King Jr. said in 1958, "the ultimate responsibility for [the Till lynching] and other tragic events must rest with the American people themselves."[22]

We are still killing black youth because we have not yet killed white supremacy. As a political program white supremacy avers that white peo-ple have a right to rule. That is obviously morally unacceptable, and few of its devotees will speak its name. But that enfeebled faith is not nearly so insidious and lethal as its robust, covert, and often unconscious cousin: the assumption that God has created humanity in a hierarchy of moral, cultural, and intellectual worth, with lighter-skinned people at the top and darker-skinned people at the bottom. Unfortunately this poisonous notion is as dangerous in the minds of people of color as it is in the minds of whites. "The glorification of one race and the consequent debasement of another—or others—always has been and always will be a recipe for murder," writes James Baldwin.[23] It also remains a recipe for toxic self-hatred.

The ancient lie remains lethal. It shoots first and dodges questions later. White supremacy leaves almost half of all African American children growing up in poverty in a de-industrialized urban wasteland. It abandons the moral and practical truth embodied in *Brown v. Board of Education* and accepts school resegregation even though it is poisonous to the poor. Internalized white supremacy in the minds of black youth guns down other black youth, who learn from media images of themselves that their lives are worth little enough to pour out in battles over street corners. White supremacy also trembles the hands of some law enforcement officers and vigilantes who seem unable to distinguish between genuine danger and centuries-old phantoms.

To see beyond the ghosts, all of us must develop the moral vision and political will to crush white supremacy—both the political program and the concealed assumptions. We have to come to grips with our own history—not only genocide, slavery, exploitation, and systems of oppression, but also the legacies of those who resisted and fought back and still fight back. We must find what Dr. King called the "strength to love." New social movements must confront head-on the racial chasm in American life. "Not everything that is faced can be changed," Baldwin instructs, "but nothing can be changed until it is faced."[24]

Our strivings will unfold in a fallen world, among imperfect people who have inherited a deeply tragic history. There will be no guarantee of success. But we have guiding spirits who still walk among us. We have the courtroom of historical memory, where Reverend Moses Wright still stands and says, "There he is." We have the boundless moral landscape where Mamie Bradley still shakes the earth with her candor and courage. We have the bold voices of the Black Lives Matter movement, demanding justice now and reminding us to remember Emmett Till, to say his name. We have the enduring NAACP and the interracial "Moral Mondays" coalition spreading out of North Carolina, like the sit-ins once did, and dozens of other similar crusades across the country.[25] We can still hear the marching feet of millions in the streets of America, all of them belonging to the children of Emmett Till.

ACKNOWLEDGMENTS

In the dark mines of this story, I was grateful that I never worked alone. My first debts are to David Cecelski, William H. Chafe, Steve Kantrowitz, and Craig Werner, unwavering friends and brilliant editors who lived with this book for years. Likewise, historians John Dittmer, Danielle McGuire, Lane Windham, Curtis Austin, Christopher Metress, David Beito, and Jane Dailey all offered vital comments. The brilliant Evan Lewis gave me priceless help and friendship, as did her parents, Ken Lewis and Holly Ewell-Lewis.

Other scholars and writers—Dan Carter, Will Jones, Kevin Kruse, Charles McKinney, Jerry Mitchell, Adriane Lentz-Smith, and Jason Morgan Ward—all offered crucial aid.

My research assistant, Melody Ivins, helped dredge up and digest much of this story and offered years of constant encouragement. She and Wilmarie Cintron-Muniz, Michael Grathwohl, Sam Tyson, and Vernon Tyson helped ransack the Mississippi archives, as did Simon Balto and Amanda Klonsky in the Chicago Public Library.

I am thankful to my colleagues at the Center for Documentary Studies at Duke University. Wesley Hogan sat with me by the cornfield and shared her brilliant insights. Tom Rankin taught me how to cook a pig and escorted me to the ruins of Bryant's Grocery and Meat Market. Mike Wiley's *Dar He*, a dramatic rendition of the Emmett Till story, fired my imagination, particularly when I watched him perform it at a refurbished tractor dealership not far from where Emmett Till died. Mary D. Williams inspired, comforted, and carried me in our public work and personal friendship. Jennifer Dixon-McKnight, Theo Luebke,

Will Griffin, and Sarah Rogers kept our classes on track so that I could remain preoccupied with Emmett Till.

Other friends also kept me in my right mind. I am grateful to Rev. Dr. William J. Barber, II; Herman Bennett; Nick Biddle; Sam Bridges; Vera Cecelski; Lorna Chafe; Katherine Charron; Louise and Steve Coggins; Jim Conway; Mary Ellen Curtin; Suzanne Desan; Kirsten Fischer; Barbara Forrest; Christina Greene; Laura Hanson; Pernille Ipsen; Rhonda Lee; Eddie McCoy; Lettie McCoy; Al McSurely; Jennifer Morgan; Leslee Nelson; Drew Ross; Rob Stephens; Doug Tanner; and Gayle Weitz.

My parents, Vernon and Martha Tyson, have been my bright and morning stars, as have sisters Boo, Julie, and Lori. The Morgans of Corapeake have inspired, comforted, and endured me for decades. Special thanks to Susan Evans for the author photograph. My brother, Vern Tyson, sets a fine example of how to embrace life even when it cuts you, and to him this book is dedicated with love.

I am grateful to the folks at Simon & Schuster, including my fine editor, Priscilla Painton, and also Sophia Jimenez, Megan Hogan, and Amanda Lang as well as copyeditors Navorn Johnson and Judith Hoover. Thomas LeBien gave matchless guidance and friendship at every stage. I wish I could write a hundred more books for my brilliant agent, Charlotte Sheedy, who has been a steadfast warrior and a stalwart friend.

My family is the heart of whatever I accomplish. My daughter, Hope, shines in my heart and in the world, becoming more compelling and clear-eyed with every passing year. And I can only admire my son, Sam, for the way he carries himself and others, building friendships, furniture, songs, and communities with equal facility. I love them both dearly. Their mother, Perri Morgan, is the pearl beyond price. Her love is not just something she feels but something she does, and I am beyond lucky to enjoy it. I am grateful to her for helping to give me this wonderful life, precious and all too short, which I hope to enjoy at her side for many more years.

NOTES

1: NOTHING THAT BOY DID

1 Carolyn Bryant Donham, interview by the author, Raleigh, NC, September 8, 2008; accompanying handwritten notes by the author, Timothy B. Tyson Papers, Southern Historical Collection, University of North Carolina at Chapel Hill, closed until 2038 by her request (hereafter Bryant, interview). Unless otherwise specified, all Carolyn Bryant quotations are from this interview.

2 Quoted in "*L'affaire Till* in the French Press," *Crisis*, December 1955, 601.

3 Interview with William Bradford Huie, by Blackside, Inc., August 1979, for *Eyes on the Prize: America's Civil Rights Years (1954–1975)*, Washington University Libraries, Film and Media Archive, Henry Hampton Collection, accessed March 31, 2016, http://digital.wustl.edu/cgi/t/text/text-idx?c=eop;cc=eop;rgn=main;view=text;idno=hui0015.1034.050 (hereafter Huie interview).

4 Timothy B. Tyson, *Blood Done Sign My Name* (New York: Crown, 2004).

5 William Bradford Huie, "Shocking Story of Approved Killing in Mississippi," *Look*, January 24, 1956; Bob Ward, "William Bradford Huie Paid for Their Sins," *Writer's Digest*, September 1974, 16–22.

6 Devery S. Anderson, *Emmett Till: The Murder That Shocked the World and Propelled the Civil Rights Movement* (Jackson: University Press of Mississippi, 2015), 324–25.

7 Carolyn Bryant testimony, *State of Mississippi vs. J. W. Milam and Roy Bryant, In the Circuit Court Second District of Tallahatchie County,*

Seventeenth Judicial District, State of Mississippi, September Term, 1955, transcript (hereafter trial transcript), 268–72.

8 "Still Heaping Criticism on Mississippi," *Greenwood Morning Star*, November 6, 1955.

9 Carolyn Bryant testimony, trial transcript, 272–74.

10 Dan Wakefield, "Justice in Sumner," *Nation*, October 1, 1955, in Christopher Metress, ed., *The Lynching of Emmett Till: A Documentary Narrative* (Charlottesville: University of Virginia Press, 2002), 120–24.

11 U.S. State Department, *Treatment of Minorities in the United States: Impact on Our Foreign Relations, Part A: A Summary Review*, RG 59 811.411/4-1956, box 4158, National Archives.

12 See, for example, Clenora Hudson-Weems, *Emmett Till: The Sacrificial Lamb of the Civil Rights Movement* (Bloomington, IN: AuthorHouse, 2006). See also Anderson, *Emmett Till*.

13 Mamie Till-Mobley and Christopher Benson, *Death of Innocence: The Story of the Hate Crime That Changed America* (New York: One World/Ballantine, 2004).

14 Carolyn Bryant Donham with Marsha Bryant, "More than a Wolf Whistle: The Story of Carolyn Bryant Donham," unpublished memoir, Timothy B. Tyson Papers, Southern Historical Collection, University of North Carolina at Chapel Hill, closed until 2038 by her request (hereafter Bryant, "Wolf Whistle").

15 Attorneys' notes of interview with Carolyn Bryant, August 30, 1955. Thanks to Jerry Mitchell at the *Jackson Clarion-Ledger* for sharing this document.

2: BOOTS ON THE PORCH

1 Although many sources refer to Reverend Wright as "Mose," Wright's son Simeon reveals that his proper name was Moses and that "Mose" was a nickname. Therefore I have used "Moses" throughout except in quotations. Simeon Wright with Herb Boyd, *Simeon's Story: An Eyewitness Account of the Kidnapping of Emmett Till* (Chicago: Lawrence Hill Books, 2010), dedication, 15.

2 *Chicago Tribune*, September 19, 1955; Wright, *Simeon's Story*, 32.

3 Robert Denley, "Kinsman Recalls Tragic Night on Eve of Trial," *Chicago Defender*, September 24, 1955; Moses Wright testimony, trial transcript, 46; Wright, *Simeon's Story*, 25–36, 55.

4 "Kin Tell How Murdered Boy Was Abducted," *Chicago Daily Tribune*,

September 3, 1955. For Elizabeth Wright's intention to wake Emmett and slip him out the back door, see also "Newspapers Over State Blast Murder of Negro," *Jackson Daily News*, September 3, 1955.

5   Wright, *Simeon's Story*, 55–60.

6   Moses Wright testimony, trial transcript, 39.

7   Wright, *Simeon's Story*, 55, 57.

8   Denley, "Kinsman Recalls Tragic Night on the Eve of Trial." Moses Wright testified in court that Emmett was in the bed with Simeon. Moses Wright testimony, trial transcript, 32–33.

9   Wright, *Simeon's Story*, 37–39, 56.

10 Paul Holmes, "Uncle Tells How Kidnapers Invaded Home and Seized Till," *Chicago Daily Tribune*, September 19, 1955.

11 Moses Wright testimony, trial transcript, 12; Denley, "Kinsman Recalls Tragic Night on Eve of Trial."

12 Moses Wright testimony, trial transcript, 36; Denley, "Kinsman Recalls Tragic Night on Eve of Trial."

13 Federal Bureau of Investigation Prosecutive Report Concerning _____ ___, 87 (hereafter FBI Report), https://vault.fbi.gov/Emmett%20Till%20/Emmett%20Till%20Part%2001%20of%2002/view, accessed June 21, 2016. Crosby Smith reported that Milam and Bryant were "plenty drunk." See David A. Shoshtak, "Crosby Smith: Forgotten Witness to a Mississippi Nightmare," *Negro History Bulletin*, December 1974, 322.

14 Holmes, "Uncle Tells How Kidnapers Invaded Home and Seized Till."

15 Moses Wright testimony, trial transcript, 38. See also FBI Report, 53.

16 Anderson, *Emmett Till*, 370, conjectures that the group that kidnapped Till comprised of J. W. Milam, Roy Bryant, Carolyn Bryant, and Henry Lee Loggins. It is impossible to say with certainty.

17 Amos Dixon, "Milam Master-Minded Emmett Till Killing," *California Eagle*, February 2, 1956; Denley, "Kinsman Recalls Tragic Night on Eve of Trial."

18 Quoted in Stanley Nelson, producer and director, *The Murder of Emmett Till*, 2003, *American Experience*, PBS, transcript, accessed April 1, 2016, http://www.pbs.org/wgbh/amex/till/filmore/pt.html.

19 Moses Wright testimony, trial transcript, 34–35.

20 "Events Night of Kidnaping Told by Slain Boy's Cousin," *Jackson Daily News*, September 1, 1955.

21 "Kin Tell How Murdered Boy Was Abducted," *Chicago Daily Tribune*, September 3, 1955.

22 Lea Thomas, "The Day That Emmett Died," *Jackson Free Press*, November 30, 2005; Anderson, *Emmett Till*, 36–37. See also "Newspapers over State Blast Murder of Negro."

23 "Slain Boy's Kinfolk Tell of Begging White Men to Let Him Off with Whipping," *Jackson Daily News*, September 3, 1955; Dixon, "Milam Master-Minded Emmett Till Killing."

24 Moses Wright testimony, trial transcript, 19.

25 Denley, "Kinsman Recalls Tragic Night on Eve of Trial." Reverend Wright acknowledged in court that he could not see whether the vehicle was a car or a pickup truck. See James Featherston, "Slain Boy's Uncle Points Finger at Bryant, Milam, But Admits Light Was Dim," *Jackson Daily News*, September 21, 1955. Stephen Whitfield claims that Wright "saw a fourth person [in the car], a woman (presumably Carolyn Bryant)." But Wright made no such claim in court or elsewhere, explaining that he could not see the people in the vehicle. Whitfield cites "'Murder,' White Says, Promises Prosecution," *Chicago Defender*, September 10, 1955, but the article contains no such claim. See Stephen J. Whitfield, *A Death in the Delta: The Story of Emmett Till* (Baltimore: Johns Hopkins University Press, 1988), 55, 160n13. For the vehicle pulling away without headlights and Wright unable to see it, see Moses Wright testimony, trial transcript, 46–48.

3: GROWING UP BLACK IN CHICAGO

1 Till-Mobley and Benson, *Death of Innocence*, 98–99. It is not clear whether Wright was still an active minister at the church in East Money. He may or may not have retired but still attended, and it stands to reason that he would preach at least occasionally.

2 Olive Arnold Adams, "Time Bomb: Mississippi Exposed and the Full Story of Emmett Till," in Metress, ed., *The Lynching of Emmett Till*, 213–24.

3 Till-Mobley and Benson, *Death of Innocence*, 98–100. Throughout the book, I call the boy "Emmett" in order to paint him as a sympathetic human being, someone with whom we become fairly intimate in the course of things, and someone who has typically been portrayed as an icon rather than a person. I do the same with his mother and also with Carolyn Bryant, and for the same reasons. No disrespect is intended.

4 Carl Hirsch, "This Was Emmett Louis Till," *Daily Worker*, October 9, 1955.

5 Chicago Demographics in 1950 map, http://www.bing.com/images/search?q=chicago_demographics_in_1950_map.jpg&qpvt=Chicago_Demographics

_in_1950_map.jpg&qpvt=Chicago_Demographics_in_1950_map
.jpg&qpvt=Chicago_Demographics_in_1950_map.jpg&FORM=IGRE,
accessed June 21, 2016.

6   Adam Cohen and Elizabeth Taylor, *American Pharaoh: Richard J. Daley:
    His Battle for Chicago and the Nation* (New York: Little, Brown, 2000),
    16–18, 29.

7   Hirsch, "This Was Emmett Louis Till."

8   Rick Swaine, *The Integration of Major League Baseball*, reprint edition
    (Jefferson, NC: McFarland, 2012), 34–43.

9   "Untold Story," *Chicago Daily Defender*, March 5, 1956; Till-Mobley and
    Benson, *Death of Innocence*, 71–72.

10 Isabel Wilkerson, *The Warmth of Other Suns: The Epic Story of America's
    Great Migration* (New York: Random House, 2010), 9, 268, 352. The de-
    finitive work for black Southern migration to Chicago is James Grossman,
    Land of Hope: Chicago, Black Southerners, and the Great Migration (Chi-
    cago: University of Chicago Press, 1989).

11 Beryl Satter, *Family Properties: How the Struggle over Race and Real Es-
    tate Transformed Chicago and Urban America* (New York: Picador, 2010),
    39; Grossman, *Land of Hope*, 174.

12 William Tuttle, *Race Riot: Chicago in the Red Summer of 1919* (New
    York: Atheneum, 1970), 3–10.

13 Davarian Baldwin, *Chicago's New Negroes: Modernity, the Great Migra-
    tion, and Black Urban Life* (Chapel Hill: University of North Carolina
    Press, 2007), 7.

14 Adriane Lentz-Smith, *Freedom Struggles: African Americans and World
    War I* (Cambridge, MA: Harvard University Press, 2011).

15 W. E. B. Du Bois, "Returning Soldiers," *Crisis* 18 (May 1919): 13.

16 Tuttle, *Race Riot*, 212.

17 For Marcus Garvey, see Edmund David Cronon, *Black Moses: The Story of
    Marcus Garvey and the Universal Negro Improvement Association* (Madi-
    son: University of Wisconsin Press, 1955); Tony Martin, *Race First: The
    Ideological and Organizational Struggle of Marcus Garvey and the Univer-
    sal Negro Improvement Association* (Dover, MA: Majority Press, 1976);
    Judith Stein, *The World of Marcus Garvey* (Baton Rouge: Louisiana State
    University Press, 1986); Mary G. Rollinson, *Grassroots Garveyism: The
    Universal Negro Improvement Association in the Rural South, 1920–1927*
    (Chapel Hill: University of North Carolina Press, 2007). See also Rob-
    ert A. Hill, ed., *The Marcus Garvey and Universal Negro Improvement*

Association Papers, 11 vols. (Berkeley: University of California Press, 1983–2006; Durham, NC: Duke University Press, 2011).

18 Tuttle, *Race Riot*, 222–26.

19 Grossman, *Land of Hope*, 74–77.

20 Faith Holsaert et al., eds., *Hands on the Freedom Plow: Personal Accounts by Women in SNCC* (Urbana: University of Illinois Press, 2012), 63.

21 Gilbert R. Mason and James Patterson Smith, *Beaches, Blood, and Ballots: A Black Doctor's Civil Rights Struggle* (Jackson: University Press of Mississippi, 2007), 19.

22 Wilkerson, *The Warmth of Other Suns*, 275.

23 Craig Werner, *Higher Ground: Stevie Wonder, Aretha Franklin, Curtis Mayfield, and the Rise and Fall of American Soul* (New York: Crown, 2004), 33.

24 Satter, *Family Properties*, 40–47; Werner, *Higher Ground*, 33.

25 Adam Green, *Selling the Race: Culture, Community, and Black Chicago, 1940–1955* (Chicago: University of Chicago Press, 2007), 183.

26 Werner, *Higher Ground*, 31.

27 For a fuller examination of the Trumbull Park Homes fiasco, see Chicago, Mayor's Commission on Human Relations, *The Trumbull Park Homes Disturbances: A Chronological Report, August 4, 1953 to June 30, 1955*, 1955; Howard Mayhew, *Racial Terror at Trumbull Park* (New York: Pioneer Press, 1954). Frank London Brown, *Trumbull Park* (1959; Lebanon, NH: University Press of New England, 2005), is a novel that conveys a sense of the experience of the black families.

28 Green, *Selling the Race*, 185–91; Cohen and Taylor, *American Pharaoh*, 1014.

29 Cohen and Taylor, *American Pharaoh*, 102–3, 175.

30 Green, *Selling the Race*, 181.

31 Cohen and Taylor, *American Pharaoh*, 58, 92–96, 128–29.

32 Green, *Selling the Race*, 181; Cohen and Taylor, *American Pharaoh*, 95; Werner, *Higher Ground*, 32.

33 Werner, *Higher Ground*, 32; Cohen and Taylor, *American Pharaoh*, 96.

34 Cohen and Taylor, *American Pharaoh*, 7–12, 16, 21, 33, 45, 57–58, 125–26, 134–35, 172–74.

4: EMMETT IN CHICAGO AND "LITTLE MISSISSIPPI"

1 Till-Mobley and Benson, *Death of Innocence*, 18–19, 22.

2 Ibid., 19. Neil R. McMillen, *Dark Journey: Black Mississippians in the Age of Jim Crow* (Urbana: University of Illinois Press, 1989), 252, reports that

large numbers of lynchings in Mississippi went unreported in any newspaper.

3 McMillen, *Dark Journey*, 229–30. See also Jacquelyn Dowd Hall, *Revolt against Chivalry: Jessie Daniel Ames and the Women's Campaign against Lynching* (New York: Columbia University Press, 1993), 134–35.

4 Richard Wright, *Black Boy* (1945; New York: Harper Perennial, 1993), 172.

5 Till-Mobley and Benson, *Death of Innocence*, 19.

6 Ibid., 3.

7 Gerald V. Stokes, *A White Hat in Argo: Family Secrets* (Lincoln, NE: iUniverse, 2004), 91–92.

8 Ibid., 86–87.

9 Till-Mobley and Benson, *Death of Innocence*, 23.

10 Ibid., 21; E. Franklin Frazier, "The Negro Family in Chicago," PhD diss., University of Chicago, 1932.

11 Till-Mobley and Benson, *Death of Innocence*, 18, 21.

12 Stokes, *A White Hat in Argo*, 98.

13 "Mamie Bradley's Untold Story," *Chicago Daily Defender*, February 29, 1956; Till-Mobley and Benson, *Death of Innocence*, 3–5, 11.

14 Till-Mobley and Benson, *Death of Innocence*, 11–12.

15 "Mamie Bradley's Untold Story"; Till-Mobley and Benson, *Death of Innocence*, 3–4, 13, 16–17.

16 "Mamie Bradley's Untold Story"; Anderson, *Emmett Till*, 11–12; Till-Mobley and Benson, *Death of Innocence*, 3–4, 13, 16–17.

17 "Mamie Bradley's Untold Story."

18 Till-Mobley and Benson, *Death of Innocence*, 27.

19 Ibid., 38.

20 "Mamie Bradley's Untold Story"; Till-Mobley and Benson, *Death of Innocence*, 37–39.

21 Till-Mobley and Benson, *Death of Innocence*, 39.

22 Ibid., 30–31, 36.

23 Ibid., 46, 49–53.

24 Ibid., 56.

25 Till-Mobley and Benson, *Death of Innocence*, 56–57.

26 Ibid., 40.

27 Wright, *Simeon's Story*, 47.

28 Till-Mobley and Benson, *Death of Innocence*, 81.

29 Albert Amateau, "Chelsea Woman Led Battle to End Beach Segregation in

1960," *Villager*, August 11–17, 2011. See especially Arnold Hirsch, *Making the Second Ghetto: Race and Public Housing in Chicago, 1940–1960* (Chicago: University of Chicago Press, 1998), 63, 65.

30 Till-Mobley and Benson, *Death of Innocence*, 82.

31 Whitfield, *A Death in the Delta*, 15.

32 Werner, *Higher Ground*, 68–69.

33 Till-Mobley and Benson, *Death of Innocence*, 83.

34 John Barrow, "Here's a Picture of Emmett Till by Those Who Knew Him," *Memphis Tri-State Defender*, September 24, 1955.

35 Till-Mobley and Benson, *Death of Innocence*, 78.

36 "Mamie Bradley's Untold Story," *Chicago Daily Defender*, March 5, 1956.

37 Barrow, "Here's a Picture of Emmett Till by Those Who Knew Him."

38 Till-Mobley and Benson, *Death of Innocence*, 63–64, 94; Moses Wright testimony, trial transcript, 48.

39 Till-Mobley and Benson, *Death of Innocence*, 103–4; Adams, "Time Bomb," in Metress, ed., *The Lynching of Emmett Till*, 213–24. The quote from Moses Wright is from Sam Johnson, "State Will Not Ask Death Penalty in Trial of White Men at Sumner," *Greenwood Commonwealth*, September 19, 1955.

5: PISTOL-WHIPPING AT CHRISTMAS

1 Bryant Donham, "Wolf Whistle," 9–10.

2 Ibid., 9.

3 See Jack Temple Kirby, "The Southern Exodus, 1910–1960: A Primer for Historians," *Journal of Southern History* 49.4 (1983): 585–600. For the Great Migration, see, among others, Wilkins, *The Warmth of Other Suns*; Grossman, *Land of Hope*.

4 David M. Oshinsky, *"Worse than Slavery": Parchman Farm and the Ordeal of Jim Crow Justice* (New York: Simon & Schuster, 1996), 139.

5 Ibid., 149–50, 161. For Camp A specifically, see Paul Oliver, *Conversations with the Blues* (Cambridge, UK: Cambridge University Press, 1997), 58.

6 Oshinsky, *"Worse than Slavery,"* 151, 153.

7 Bryant, "Wolf Whistle," 13.

8 Ibid., 16–17.

9 Ibid., 17.

10 Anderson, *Emmett Till*, 24.

11 Bryant, "Wolf Whistle," 21–22.

12 James Henry Hammond was the first to expound this theory. See Drew

Gilpin Faust, *James Henry Hammond and the Old South* (Baton Rouge: Louisiana State University Press, 1982), 346–47.

13 Bryant, interview.

14 "'Were Never into Meanness,' Says Accused Men's Mother," *Memphis Commercial Appeal*, September 2, 1955, quoted in Metress, ed., *The Lynching of Emmett Till*, 34–35.

15 FBI Report, 22–23.

16 Bryant, "Wolf Whistle," 17–19, 20, 22.

17 Ibid., 27.

18 Adams, "Time Bomb," in Metress, ed., *The Lynching of Emmett Till*, 213–24.

19 FBI Report, 32.

20 Adams, "Time Bomb," in Metress, ed., *The Lynching of Emmett Till*, 213–24.

21 Bryant, interview.

22 William Bradford Huie, "Shocking Story of Approved Killing in Mississippi," *Look*, January 24, 1956. See also FBI Report, 25–26; "Milam Is Pictured as a War Hero Who Also Snatched Negro from Drowning," *Jackson Daily News*, September 20, 1955.

23 FBI Report, 25–26.

24 "Milam Is Pictured as a War Hero Who Also Snatched Negro from Drowning."

25 "Mother of Pair Accused in Delta Slaying 'Will Stand by Them,'" *Jackson Daily News*, September 2, 1955; Huie quoted in Whitaker, "A Case Study," 108–9.

26 FBI Report, 108.

27 Quoted in Ronald Turner, "Remembering Emmett Till," *Howard Law Journal* 38 (1994–95): 422.

28 Bryant, interview.

29 Ibid.; FBI Report, 13, 25–26.

30 Huie, "Shocking Story of Approved Killing in Mississippi."

6: THE INCIDENT

1 Moses Wright testimony, trial transcript, 58–59; Anderson, *Emmett Till*, 27; "Mother Waits in Vain for Her 'Bo,'" *Chicago Defender*, September 10, 1955, in Metress, ed., *The Lynching of Emmett Till*, 30–31; Herb Boyd, "The Real Deal on Emmett Till," *Black World Today*, May 18, 2004, accessed April 5, 2016, http://www.afro-netizen.com/2004/05/the_real_deal_o.html. For information on each of these young people, see Devery Anderson's website, www.emmetttillmurder.com, under "Who's Who in the Emmett Till Story."

2 Bryant, "Wolf Whistle," 31.

3 Till-Mobley and Benson, *Death of Innocence*, 102.

4 "Kidnapped Boy Whistled at Woman," *Chicago Daily Tribune*, August 30, 1955.

5 Wright, *Simeon's Story*, 50.

6 "Mrs. Roy Bryant, 9-2-55," lawyer's notes, copy in possession of the author. My thanks to reporter Jerry Mitchell for sharing these papers.

7 "Bryant and Milam Face Murder Charge in the Slaying of 15 Year Old Negro Boy," *Greenwood Morning Star*, September 1, 1955. This "good-bye" seems to be a point of considerable agreement on both sides. Maurice Wright, too, reported that on his way out of the store Emmett turned and told Carolyn "Goodbye." See "Events Night of Kidnapping Told by Slain Boy's Cousin," *Jackson Daily News*, September 1, 1955.

8 Clark Porteous, "Grand Jury to Get Case of Slain Negro Boy Monday," *Memphis Press-Scimitar*, September 1, 1955.

9 "Nation Shocked, Vow Action in Lynching of Chicago Youth," *Baltimore Afro-American*, September 10, 1955.

10 Keith Beauchamp, "The Murder of Emmett Louis Till: The Spark That Started the Civil Rights Movement," n.d., http://www.thefreelibrary.com /The+murder+of+Emmett+Louis+Till%3A+the+spark+that+started+the +Civil. . .-a0131248351.

11 Philip Dray, *At the Hands of Persons Unknown: The Lynching of Black America* (New York: Random House, 2002), 423. I take the phrase from W. E. B. Du Bois's introduction to the 1953 Jubilee edition of *The Souls of Black Folk*, but it was coined by Walter Bagehot in his 1872 *Physics and Politics* and was common enough in contemporary scholarly parlance for Du Bois to employ without attribution.

12 Thomas, "The Day That Emmett Till Died."

13 Wright, *Simeon's Story*, 51.

14 Mattie S. Colon and Robert Elliott, "Mother Waits in Vain for Her 'Bo,'" *Memphis Times-Scimitar*, September 10, 1955.

15 Boyd, "The Real Deal on Emmett Till."

16 Wright, *Simeon's Story*, 51.

17 Ibid., 52–53. Wheeler Parker also reported back in September 1955, "A girl told us we hadn't heard the last of it." See Barrow, "Here's a Picture of Emmett Till by Those Who Knew Him."

7: ON THE THIRD DAY

1 "Uncle Tells How 3 Kidnapers Invaded Home and Seized Till," part 1, *Chicago Daily Tribune*, September 19, 1955. See also Dixon, "Milam Master-Minded Emmett Till Killing."

2 "Urges Husband to Leave Dixie," *Chicago Defender*, September 17, 1955.

3 Shoshtak, "Crosby Smith," 320–25; Bryant, "Wolf Whistle," 44.

4 Till-Mobley and Benson, *Death of Innocence*, 116–18.

5 Rayfield Mooty, Elizabeth Balanoff Labor Oral History Collection, Roosevelt University, transcript, 1984, 140–41, accessed April 1, 2016, www.roosevelt.edu/~/media/Files/pdfs/Library/OralHistory/38-Mooty (hereafter Mooty, interview).

6 "Bryant and Milam Face Murder Charge in Slaying of 15 Year Old Negro Boy," *Greenwood Morning Star*, September 1, 1955; *Greenwood Commonwealth*, August 30, 1955.

7 Shoshtak, "Crosby Smith," 320–25.

8 Dixon, "Till Case: Torture and Murder," *California Eagle*, February 16, 1956.

9 Bryant, "Wolf Whistle," 46–47.

10 George Smith testimony, trial transcript, 119–20. Sheriff Smith also confirmed that the issue was alleged "ugly remarks" ("White Storekeeper Held in Abduction of Negro Youth," *Jackson Daily News*, August 29, 1955). See also "'Kidnaped' Negro Boy Still Missing, Fear Foul Play," *Jackson Daily News*, August 30, 1955.

11 Anderson, *Emmett Till*, 42–43.

12 Robert Hodges testimony, trial transcript, 103–4. For a description of Hodges, see John Herbers, "Contradictions Develop as Testimony in Till Trial Begins," *Greenwood Morning Star*, September 22, 1955.

13 FBI Report, 80; Bryant, interview.

14 "No Clues in Disappearance of Negro Youth," *Greenwood Morning Star*, August 31, 1955.

15 Darryl Christopher Mace, "Regional Identities and Racial Messages: The Print Media's Stories of Emmett Till," PhD diss., Temple University, 2007, 120.

16 "Boy's Slaying Held Murder by Gov. White," *Chicago Daily Tribune*, September 2, 1955.

17 "Search Halted for Woman in Negro Slaying," *Birmingham News*, September 3, 1955.

18 Bryant, "Wolf Whistle," 47.

19 "Bryant's Brother Claims Charges Are All 'Politics,'" *Memphis Commercial Appeal*, September 3, 1955.

20 "Mississippi Sheriff Voices Doubt Body That of Till," *Greenwood Morning Star*, September 4, 1955.

21 Nelson, *The Murder of Emmett Till*.

22 B. L. Mims testimony, trial transcript, 113–18; Robert Hodges testimony, trial transcript, 103–12.

23 James Featherston, "White Orders Investigation in Slaying of Delta Negro: White Deplores Slaying in a Note to NAACP Which Is Creating a National Issue," *Jackson Daily News*, September 1, 1955. See also *Chicago Tribune*, September 1, 1955; *Memphis Commercial Appeal*, September 1, 1955.

24 *Greenwood Commonwealth*, August 31, 1955; Chester Miller testimony, trial transcript, 67–82.

25 Moses Wright testimony, trial transcript, 51–52; Chester Miller testimony, trial transcript, 70–71, 73–74, 97–99. For Wright's identification of the body, see also *Arkansas Gazette*, September 4, 1955.

26 Chester Miller testimony, trial transcript, 70–71, 73–74, 97–99.

27 "Muddy River Gives Up Body of Brutally Slain Negro Boy," *Memphis Commercial Appeal*, September 1, 1955.

28 *New York Post*, September 1, 1955; Featherston, "White Orders Investigation in Slaying of Delta Negro."

29 Till-Mobley and Benson, *Death of Innocence*, 130.

30 Joe Atkins, "Chicago Youth Was 'Sacrificial Lamb,'" *Jackson Daily News*, August 25, 1985, clipping, box 90, folder 18, Emmett Till folder (1955–1985), Erle Johnston Papers, University of Southern Mississippi.

31 Suzanne E. Smith, *To Serve the Living* (Cambridge, MA: Belknap Press of Harvard University Press, 2010), 126; Whitaker, "A Case Study in Southern Justice," 118–19; C. M. Nelson testimony, trial transcript, 182.

32 Till-Mobley and Benson, *Death of Innocence*, xxii.

33 *Pittsburgh Courier* quoted in Mace, "Regional Identities and Racial Messages," 169.

34 Charles Frederick Weller to Anne Braden, January 31, 1956, box 26, folder 6, Carl and Anne Braden Papers, Wisconsin Historical Society.

8: MAMA MADE THE EARTH TREMBLE

1 Till-Mobley and Benson, *Death of Innocence*, 130–31.

2 Mattie Smith Colin, "Mother's Tears Greet Son Who Died a Martyr," *Chicago Defender*, September 10, 1955.

3 Mace, "Regional Identities and Racial Messages," 219, quotes the coverage in the *Chicago Sun-Times*.

4 Till-Mobley and Benson, *Death of Innocence*, 132. See also Smith, *To Serve the Living*, 127. For a good sense of the gender dynamics of her public performances, see Ruth Feldstein, "'I Wanted the Whole World to See': Race, Gender and Constructions of Motherhood in the Death of Emmett Till," in Joanne Meyerowitz, ed., *Not June Cleaver: Women and Gender in Postwar America, 1945–1960* (Philadelphia: Temple University Press, 1993), 263–303.

5 Mooty, interview.

6 Colin, "Mother's Tears Greet Son Who Died a Martyr."

7 Charles M. Payne, *I've Got the Light of Freedom: The Organizing Tradition and the Mississippi Freedom Struggle* (Berkeley: University of California Press, 1995), 13–14. See also Langston Hughes, "Langston Hughes Wonders Why No Lynching Probes," *Chicago Defender*, October 1, 1955, in Metress, ed., *The Lynching of Emmett Till*, 124–27; "2 Negro Boys Lynched," *New York Times*, October 13, 1942. For a full and brilliant account of this lynching that lets it speak to a century of racial violence in the United States, see Jason Ward, *Hanging Bridge: Racial Violence and America's Civil Rights Century* (New York: Oxford University Press, 2016).

8 Aaron Henry interview by John Dittmer and John Jones, April 22, 1981, transcript, 7–8, Mississippi Department of Archives and History; Erle Johnston, *Mississippi's Defiant Years, 1953–1973: An Interpretive History with Personal Experiences* (Forest, MS: Lake Harbor, 1990), 35.

9 *Pittsburgh Courier*, September 17, 1955, 4.

10 Mooty, interview.

11 Till-Mobley and Benson, *Death of Innocence*, 131–32.

12 Wil Haygood, "The Man from *Jet*: Simeon Booker Not Only Covered a Tumultuous Era, He Lived It," *Washington Post*, July 15, 2007.

13 Till-Mobley and Benson, *Death of Innocence*, 133.

14 Ibid., 132–37. For Gene Mobley Jr. identification of Till by his haircut, see also *Chicago Tribune*, September 4, 1955, 2.

15 *Chicago Sun-Times* quoted in Mace, "Regional Identities and Racial Messages," 220; Mooty, interview.

16 Till-Mobley and Benson, *Death of Innocence*, 139–40.

17 Whitfield, *A Death in the Delta*, 23, estimates turnout that first night at "perhaps ten thousand." *Chicago Defender*, September 10, 1955; Smith, *To Serve the Living*, 127–28, also estimates crowds of fifty thousand.

18 Michael Vinson Williams, *Medgar Evers: Mississippi Martyr* (Fayetteville: University of Arkansas Press, 2011), 126.

19 Till-Mobley and Benson, *Death of Innocence*, 139.

20 *Chicago Tribune*, September 4, 1955.

21 Wilkerson, *The Warmth of Other Suns*, 369–70.

22 Till-Mobley and Benson, *Death of Innocence*, 141, 144.

23 David Smothers, "Killing of Boy in Mississippi called 'Atrocity,'" *Jackson Daily News*, September 3, 1955.

24 Till-Mobley and Benson, *Death of Innocence*, 143.

25 *Chicago Tribune*, September 7, 1955.

26 *Chicago Tribune*, September 7, 1955; *Chicago Defender*, September 17, 1955.

27 "Milam and Bryant to Be Tried Sept. 19," *Greenwood Morning Star*, September 10, 1955.

28 *Jet*, September 15, 1955, 6–9; *Chicago Defender*, September 17, 1955.

29 Green, *Selling the Race*, 198.

30 Julian Bond, "The Media and the Movement: Looking Back from the Southern Front," in Brian Ward, ed., *Media, Culture and the Modern African American Freedom Struggle* (Gainesville: University Press of Florida, 2001), 27.

31 Ann Marie Tabb, "Perspectives in Journalism: Covering the Emmett Till Trial," MA thesis, University of Southern Mississippi, 2001, 20.

32 Green, *Selling the Race*, 180–82, 208.

9: WARRING REGIMENTS OF MISSISSIPPI

1 Tom P. Brady, *Black Monday: Segregation or Amalgamation. America Has Its Choice* (Winona, MS: Association of Citizens' Councils, 1954), 63–64. See also Hodding Carter III, *The South Strikes Back* (Garden City, NY: Doubleday, 1959), 29.

2 J. W. Milam, quoted in Huie, "Shocking Story of Approved Killing in Mississippi," 50.

3 See Douglas R. Edgerton, *The Wars of Reconstruction: The Brief, Violent History of America's Most Progressive Era* (New York: Bloomsbury, 2014); LeRae Sikes Umfleet, *A Day of Blood: The 1898 Wilmington Race Riot* (Raleigh: North Carolina Office of Archives and History, 2009); David S. Cecelski and Timothy B. Tyson, *Democracy Betrayed: The Wilmington Race Riot and Its Legacy* (Chapel Hill: University of North Carolina Press, 1998); H. Leon Prather, *We Have Taken a City: Wilmington Coup and*

Massacre of 1898 (Rutherford, NJ: Fairleigh Dickinson University Press, 1984); Helen G. Edmonds, *The Negro and Fusion Politics in North Carolina, 1894–1901* (Chapel Hill: University of North Carolina Press, 1951); David Fort Godshalk, *Veiled Visions: The 1906 Atlanta Race Riot and the Reshaping of American Race Relations* (Chapel Hill: University of North Carolina Press, 2005); Roberta Senechal, *The Sociogenesis of a Race Riot: Springfield, Illinois, in 1908* (Urbana: University of Illinois Press, 1990); Elliott Rudwick, *Race Riot at East St. Louis, July 2, 1917* (Urbana: University of Illinois Press, 1982); Nan Elizabeth Woodruff, *American Congo: The African American Freedom Struggle in the Delta* (Cambridge, MA: Harvard University Press, 2003), 74–109; Scott Ellsworth, *Death in a Promised Land: The Tulsa Race Riot of 1921* (1982; Baton Rouge: Louisiana State University Press, 1992).

4   William Bradford Huie, "Wolf Whistle," in Metress, ed., *The Lynching of Emmett Till,* 241.

5   Jason Morgan Ward, *Defending White Democracy: The Making of a Segregationist Movement and the Remaking of Racial Politics, 1936–1965* (Chapel Hill: University of North Carolina Press, 2011), 6–7.

6   Woodruff, *American Congo,* 213–14, 222.

7   Richard Wright, *12 Million Black Voices* (1941; New York: Basic Books, 2011), 11.

8   Kari Frederickson, *The Dixiecrat Revolt and the End of the Solid South, 1932–1968* (Chapel Hill: University of North Carolina Press, 2001), 2–3, 237–38.

9   Ward, *Defending White Democracy,* 102, 105.

10   Woodruff, *American Congo,* 224.

11   National Association of Post Office and General Services Employees to Amzie Moore, July 17, 1953, box 1, folder 2, Amzie Moore Papers, Wisconsin Historical Society; Amzie Moore interview with Michael Garvey, March 29 and April 13, 1977, University of Southern Mississippi Oral History, crdl.usg.edu/export/html/usm/coh/crdl_usm_coh_ohmoorea.html ?welcome (hereafter Moore interview with Garvey). For Moore's appearance, see photograph in Henry Hampton et al., eds., *Voices of Freedom: An Oral History of the Civil Rights Movement from the 1950s through the 1980s* (New York: Bantam, 1990), 139.

12   Interview with Amzie Moore, by Blackside, Inc., 1979, for *Eyes on the Prize: America's Civil Rights Years (1954–1975),* Washington University Film and Media Archive, Henry Hampton Collection, accessed April 1,

2016, http://digital.wustl.edu/cgi/t/text/text-idx?c=eop;cc=eop;rgn=main ;view=text;idno=moo0015.0109.072 (hereafter Moore, *Eyes on the Prize* interview).

13 Moore interview with Garvey.

14 Malcolm Boyd, "Survival of a Negro Leader," February 27, 1965, box 1, folder 1, Amzie Moore Papers, Wisconsin Historical Society. See also James Forman, *The Making of Black Revolutionaries* (New York: Macmillan, 1972), 278.

15 Moore interview with Garvey.

16 Ibid.

17 Charles McLaurin, interview by the author, 2012, Timothy B. Tyson Papers, Southern Historical Collection, University of North Carolina at Chapel Hill.

18 Forman, *The Making of Black Revolutionaries*, 278–79.

19 Matthew Skidmore to Amzie Moore, July 10, 1955, b. 4 f. 2, Amzie Moore Papers, Wisconsin Historical Society.

20 Matthew Skidmore to Amzie Moore, July 16, 1955, b. 1 f. 2, Amzie Moore Papers, Wisconsin Historical Society.

21 Forman, *The Making of Black Revolutionaries*, 278–79.

22 Amzie Moore interview, in Howell Raines, ed., *My Soul Is Rested: Movement Days in the Deep South Remembered* (New York: Penguin, 1977), 233.

23 Forman, *The Making of Black Revolutionaries*, 279.

24 Moore interview with Garvey.

25 Amzie Moore certificate of Honorable Discharge from the U.S. Army, January 17, 1946, Camp Shelby, Mississippi, b. 1 f. 1, Amzie Moore Papers, Wisconsin Historical Society.

26 Payne, *I've Got the Light of Freedom*, 30.

27 Forman, *The Making of Black Revolutionaries*, 279. Moore estimated that at least one returning black serviceman was killed each week for six to eight weeks. Charles Payne sensibly concludes that the figure is probably exaggerated, though life on desolate plantations, isolated and harsh, was cheap, and news of a number of killings may not have gotten out (Payne, *I've Got the Light of Freedom*, 448n6).

28 Moore interview with Garvey.

29 Moore, *Eyes on the Prize* interview.

30 John Dittmer, *Local People: The Struggle for Civil Rights in Mississippi* (Urbana: University of Illinois Press, 1994), 9.

31 Myrlie Evers-Williams and Manning Marable, eds., *The Autobiography of Medgar Evers* (New York: Basic Books, 2005), 12.

32 Williams, *Medgar Evers*, 38, 40–41; Charles Evers and Andrew Szanton, *Have No Fear: A Black Man's Fight for Respect in America* (New York: Wiley, 1997), 64.

33 Williams, *Medgar Evers*, 31, 44–51; Payne, *I've Got the Light of Freedom*, 49–50; Evers and Szanton, *Have No Fear*, 57–58. See also "Mound Bayou Man Files Application at 'U' Law School," *Jackson Daily News*, January 22, 1954.

34 Evers-Williams and Marable, *The Autobiography of Medgar Evers*, 7–9; Evers and Szanton, *Have No Fear*, 67.

35 This portrait of Howard is drawn from an excellent biography by David Beito and Linda Royster Beito, *Black Maverick: T.R.M. Howard's Fight for Civil Rights and Economic Power* (Urbana: University of Illinois Press, 2009); Dittmer, *Local People*, 32; Evers and Szanton, *Have No Fear*, 65–66; Williams, *Medgar Evers*, 56–57. For the weapons, see Beito and Beito, *Black Maverick*, xiii.

36 Jay Driskell, "Amzie Moore: The Biographical Roots of the Civil Rights Movement in Mississippi," in Susan Glisson, ed., *The Human Tradition in the Civil Rights Movement* (Lanham, MD: Rowman & Littlefield, 2006), 137.

37 "Dear Leader," Bolivar County Invitational Committee of the Proposed Delta Council of Negro Leadership, December 10, 1951, b. 7 f. 2, Amzie Moore Papers, Wisconsin Historical Society. Moore himself recalled that the first meeting took place at the H. M. Nailer School in Cleveland in 1950, but the printed invitations indicate that it was at the Cleveland Colored High School in 1951. See Forman, *The Making of Black Revolutionaries*, 279; Aaron Henry interview by John Dittmer and John Jones, April 22, 1981, transcript, 18–19, Mississippi Department of Archives and History.

38 Moore interview with Garvey.

39 "The Accomplishments and Objectives of the Regional Council of Negro Leadership of Mississippi," box 7, folder 2, Amzie Moore Papers, Wisconsin Historical Society. See also "The New Fighting South: Militant Negroes Refuse to Leave Dixie or Be Silenced," *Jet*, August 1955, 69–74.

40 Payne, *I've Got the Light of Freedom*, 31–32; Driskell, "Amzie Moore," 137–40; "The Accomplishments and Objectives of the Regional Council on Negro Leadership of Mississippi."

41 Moore interview with Garvey; Boyd, "Survival of a Negro Leader."

42 Beito and Beito, *Black Maverick*, 80; "The Accomplishments and Objectives

of the Regional Council on Negro Leadership of Mississippi"; "No Rest Room, No Gas," Regional Council on Negro Leadership press release, n.d., box A466, Group 2, NAACP Papers, Library of Congress. See also Driskell, "Amzie Moore," 139–40.

43 Evers and Szanton, *Have No Fear*, 73.

44 Myrlie Evers with William Peters, *For Us, the Living* (Garden City, NY: Doubleday, 1967), 87–88.

45 "The Accomplishments and Objectives of the Regional Council on Negro Leadership of Mississippi"; "No Rest Room, No Gas"; Beito and Beito, *Black Maverick*, 81.

46 Beito and Beito, *Black Maverick*, 79.

47 "Program of the Third Annual Meeting of the Mississippi Regional Council of Negro Leadership," May 7, 1954, Mound Bayou, Mississippi, box 7, folder 2, Amzie Moore Papers, Wisconsin Historical Society; Beito and Beito, *Black Maverick*, 88.

48 "The New Fighting South," 69. See also Akinyele Omowale Umoja, *We Will Shoot Back: Armed Resistance in the Mississippi Freedom Movement* (New York: New York University Press, 2013), 32.

49 Beito and Beito, *Black Maverick*, 82, 89.

50 Ward, *Defending White Democracy*, 65, 90, 102, 123–24.

10: BLACK MONDAY

1 Joseph Crespino, *In Search of Another Country: Mississippi and the Conservative Counterrevolution* (Princeton, NJ: Princeton University Press, 2007), 19.

2 Tom Ethridge, "Racial Crisis Spurs Sale of Book, 'Black Monday,'" *Clarion-Ledger*, August 7, 1955.

3 Neil R. McMillen, *The Citizens' Council: Organized Resistance to the Second Reconstruction, 1954–64* (Urbana: University of Illinois Press, 1971), 17–18.

4 Tom P. Brady, *Black Monday*, excerpted in Clayborne Carson et al., eds., *Eyes on the Prize Reader* (New York: Penguin, 1991), 86–89.

5 Florence Mars, *Witness in Philadelphia* (Baton Rouge: Louisiana State University Press, 1977), 53.

6 McMillen, *Citizens' Council*, 17–18.

7 Mars, *Witness in Philadelphia*, 53.

8 McMillen, *Citizens' Council*, 17.

9 Jacquelyn Dowd Hall, "The Mind That Burns in Each Body," *Southern*

Exposure, November/December 1984, 64; Hall, *Revolt against Chivalry*, 155.

10 Brady, *Black Monday*, 12.

11 Ibid., 10–11 and 10–13.

12 Ibid.

13 Ibid.

14 Wayne Addison Clark, "An Analysis of the Relationship between Anti-Communism and Segregationist Thought in the Deep South, 1948–1964," PhD diss., University of Wisconsin–Madison, 1976, 95–96.

15 Brady, *Black Monday*, excerpted in Carson et al., eds., *Eyes on the Prize Reader*, 89.

16 Clark, "An Analysis of the Relationship between Anti-Communism and Segregationist Thought in the Deep South," 84, 92, 95–96.

17 Robert Patterson interview, in Raines, *My Soul Is Rested*, 298.

18 Richard Kluger, *Simple Justice* (New York: Vintage, 1977), 279–82.

19 Ward, *Defending White Democracy*, 125.

20 Carter, *The South Strikes Back*, 22–23.

21 Phillip Luce, "Down in Mississippi—The White Citizens' Council," *Chicago Jewish Forum*, c. 1958, box 1, folder 19, Citizens' Council Papers, University of Southern Mississippi, 324.

22 McMillen, *The Citizens' Council*, 18–19; Robert Patterson interview, in Raines, *My Soul Is Rested*, 298.

23 Crespino, *In Search of Another Country*, 23.

24 Johnston, *Mississippi's Defiant Years*, 13–14.

25 Erle Johnston, "Interview with Dr. Caudill," n.d., Citizens' Council Papers, University of Southern Mississippi.

26 Phillip Abbott Luce, "The Mississippi White Citizens Council, 1954–1959," MA thesis, Ohio State University, 1960, 19.

27 FBI Report, 16–17.

28 "Pro-Segregation Groups in the South," Southern Regional Council, November 19, 1956, 1, box 3, folder 15, Citizens Council Collection, University of Southern Mississippi.

29 McMillen, *The Citizens' Council*, 18–19. For an analysis of this kind of masculinist rhetoric in the Citizens' Council movement, see Steve Estes, *I Am a Man! Race, Manhood and the Civil Rights Movement* (Chapel Hill: University of North Carolina Press, 2005), 43–48.

30 W. J. Cash, *The Mind of the South* (New York: Knopf, 1941), 117–19.

31 "Strong Organization Only Way to Combat Integration Says Patterson," *Greenwood Morning Star*, September 14, 1955.

32 Payne, *I've Got the Light of Freedom*, 35; Johnston, *Mississippi's Defiant Years*, 15–16.

33 For Lillian Smith quote, see Clark, "An Analysis of the Relationship between Anti-Communism and Segregationist Thought in the Deep South," 205.

34 Carter, *The South Strikes Back*, 49–50.

35 Dittmer, *Local People*, 45–46.

36 Payne, *I've Got the Light of Freedom*, 431.

37 "Pro-Segregation Groups in the South," 7–8.

38 Carter, *The South Strikes Back*, 17.

39 Luce, "Down in Mississippi."

40 Whitaker, "A Case Study in Southern Justice," 81.

41 "Citizens Council to Establish Official Paper at Indianola," *Greenwood Morning Star*, September 16, 1955.

42 Luce, "Down in Mississippi," 327. All issues of the *Citizen* are available at www.citizenscouncils.com.

43 Whitaker, "A Case Study in Southern Justice," 73.

44 Medgar Evers quoted in Bem Price, "Associated Press Writer Views Mississippi Problem," *Jackson Daily News*, August 21, 1955; Ruby Hurley, "News and Action," NAACP Southeast Regional Office, Birmingham, Alabama, September 1955, box 2, Medgar Evers Papers, Mississippi Department of Archives and History.

45 Dittmer, *Local People*, 46–47.

46 Price, "Associated Press Writer Views Mississippi Problem." *Jackson Daily News*, August 21, 1955.

47 "McCoy Has Reached the Limit," *Jackson Daily News*, August 22, 1955.

48 Crespino, *In Search of Another Country*, 25.

49 NAACP, "M Is for Mississippi and Murder," 1955, box 2, folder 19, Citizens' Council magazine articles, McCain Papers, University of Southern Mississippi.

50 Crespino, *In Search of Another Country*, 26.

51 This founder and financier quoted in Luce, "The Mississippi White Citizens' Council," 32n6.

52 For the White Citizens Legal Fund and the defense of Medgar Evers's murderer, see clipping from the June 1963 *Greenwood Commonwealth*, Medgar Evers FBI file, 19, Medgar Evers Papers, Mississippi Department of Archives and History.

53 Crespino, *In Search of Another Country*, 26.

54 Luce, "Down in Mississippi."

55 Ruby Hurley, Regional Secretary, to Gloster Current, Director of Branches, NAACP, New York City, box 2, folder 7, Medgar Evers Papers, Mississippi Department of Archives and History.

56 Dittmer, *Local People*, 34–36, 43.

57 Charles C. Bolton, *The Hardest Deal of All: The Battle over School Integration in Mississippi, 1870–1980* (Jackson: University Press of Mississippi, 2005), 67–69.

58 Dittmer, *Local People*, 46.

59 Carter, *The South Strikes Back*, 37.

60 Bolton, *The Hardest Deal of All*, 67.

61 Dittmer, *Local People*, 45.

62 Percy Greene quoted in Beito and Beito, *Black Maverick*, 110; "Rev. George Lee's Murderers Never Caught," *Neshoba News*, December 8, 1995; McMillen, *Citizens' Council*, 28.

63 *Jackson Daily News*, August 18, 1955; Bolton, *The Hardest Deal of All*, 73–75.

64 Medgar Evers, 1955 Annual Report from the Mississippi State Office, National Association for the Advancement of Colored People, Jackson, box 2, Medgar Evers Papers, Mississippi Department of Archives and History; Bolton, *The Hardest Deal of All*, 73–75.

65 Evers, 1955 Annual Report. For the growth of the NAACP, see Payne, *I've Got the Light of Freedom*, 40, 140.

66 Medgar Evers to Roy Wilkins, September 12, 1956, box 2, folder 9, Coleman Papers, Mississippi Department of Archives and History.

67 *Eagle Eye*, August 20, 1955, box 2, folder 7, Medgar Evers Papers, Mississippi Department of Archives and History.

68 Quoted in Steven D. Classen, *Watching Jim Crow: The Struggles over Mississippi TV, 1955–1969* (Durham, NC: Duke University Press, 2004), 32–33.

69 Quoted in Randy Sparkman, "The Murder of Emmett Till," *Slate*, June 21, 2005, http://www.slate.com/articles/news_and_politics/jurisprudence/2005/06/the_murder_of_emmett_till.html. Accessed October 31, 2016.

11: PEOPLE WE DON'T NEED AROUND HERE ANY MORE

1 Phillip Abbott Luce, "The Mississippi White Citizens Council, 1954–1959," MA thesis, Ohio State University, 1960, 3.

2 NAACP, "M Is for Mississippi and Murder," 1955, box 2, folder 19,

Citizens' Council magazine articles, McCain Papers, Special Collections, University of Southern Mississippi.

3 Aaron Henry interview by John Dittmer and John Jones, 1982, transcript, 10–11, Mississippi Department of Archives and History (hereafter cited as MDAH).

4 Jack Mendelsohn, *The Martyrs* (New York: Harper & Row, 1966), 1.

5 David Halberstam, *The Fifties* (New York: Villard, 1993), 430.

6 Mendelsohn, *The Martyrs*, 13.

7 Beito and Beito, *Black Maverick*, 112.

8 "Candidates Say Delta Negroes Aren't Democrats," *Jackson Daily News*, August 2, 1955.

9 "U.S. Won't Investigate Charges Negroes Couldn't Vote in First Primary," *Jackson Daily News*, August 16, 1955. See also Stephen Andrew Berrey, "Against the Law: Violence, Crime, State Repression, and Black Resistance in Jim Crow Mississippi," PhD diss., University of Texas at Austin, 2006, 162.

10 "Prosecution Is Sought in Negro Voting Methods," *Jackson Daily News*, August 26, 1955; "Group Named to Study Ways to Cut Down on Negro Voting: Committee Fears Negroes 'Played Too Large a Part in State's Last Elections,'" *Jackson Daily News*, August 30, 1955.

11 Margaret Price, *The Negro Voter in the South* (Atlanta: Southern Regional Council, 1957), 20.

12 Bem Price, "Associated Press Writer Views Mississippi Problem," *Jackson Daily News*, August 21, 1955; Carter, *The South Strikes Back*, 41–42.

13 Hugh Stephen Whitaker, "A Case Study in Southern Justice: The Emmett Till Case," MA thesis, Florida State University, 1963, 62, 74–75. Hereafter, Whitaker, "A Case Study."

14 Carter, *The South Strikes Back*, 41–42.

15 Photostat submitted by NAACP national executive director Roy Wilkins to U.S. Senate Judiciary Committee, May 16, 1956, in Mississippi State Sovereignty Commission online files under "Gus Courts."

16 Mendelsohn, *The Martyrs*, 7.

17 Simeon Booker, *Black Man's America* (Englewood, NJ: Prentice-Hall, 1964), 161.

18 Mendelsohn, *The Martyrs*, 2.

19 David T. Beito and Linda Royster Beito, "The Grim and Overlooked Anniversary of the Murder of the Rev. George W. Lee, Civil Rights Activist," *History News Network*, May 9, 2005, www.hnn.us/article/11744. Accessed October 31, 2016.

20 Booker, *Black Man's America*, 161–62.

21 Payne, *I've Got the Light of Freedom*, 140. For Lee's militancy, see "Is Mississippi Hushing Up a Lynching? Mississippi Gunmen Take Life of Militant Negro Minister," *Jet* 8.3 (1955): 196. Payne takes a similar view of Lee, calling him one of "the more radical leaders" in Mississippi, along with Medgar Evers, Aaron Henry, and Amzie Moore, all of whom were active in the RCNL.

22 Payne, *I've Got the Light of Freedom*, 36, 49.

23 For Wilkins's estimates of Courts and Lee's voter registration successes, see Roy Wilkins with Tom Mathews, *Standing Fast: The Autobiography of Roy Wilkins* (New York: Viking, 1982), 222. Payne, *I've Got the Light of Freedom*, 36–37, estimates "about 100" and documents the threat to sue Sheriff Shelton when he refused to accept poll tax payments. For the quote from the sheriff, see Robert A. Caro, *Master of the Senate* (New York: Knopf, 2002), 700. Courts himself estimated "about 400." "Testimony of Rev. Gus Courts, Belzoni, Miss.," in *Civil Rights—1957: Hearings before the Subcommittee on Constitutional Rights of the Committee on the Judiciary, U.S. Senate, 85th Congress, First Session*, 532 (hereafter cited as Courts testimony).

24 Roy Wilkins, testimony before U.S. Senate Judiciary Committee, May 16, 1956, in Mississippi State Sovereignty Commission online, under "Gus Courts."

25 Payne, *I've Got the Light of Freedom*, 36–37, 49. For the Citizens' Council campaign of reprisals, see http://www.splc.org/what-we-do-/civil-rights -memorial/civil-rights-martyrs/george-lee. Accessed October 31, 2016.

26 Price, "Associated Press Writer Views Mississippi Problem."

27 Anderson, *Separate but Equal*, 121; Payne, *I've Got the Light of Freedom*, 37.

28 Mendelsohn, *The Martyrs*, 3.

29 Payne, *I've Got the Light of Freedom*, 37; Mendelsohn, *The Martyrs*, 5. See also Courts testimony, 545.

30 "Rev. George Lee's Murderers Never Caught," *Neshoba News*, December 8, 2005. See also Susan Klopfer, "FBI Investigated George Lee's Murder; Suspects Never Tried," *Ezine*, December 8, 2005, http://ezinearticles .com/?FBI-Investigated-George-Lees-Murder;-Suspects-Never-Tried&id= 109869. Accessed October 31, 2016.

31 Mary Panzer, "H. C. Anderson and the Civil Rights Struggle," in Henry Clay Anderson et al., *Separate but Equal: The Mississippi Photographs of Henry Clay Anderson* (New York: PublicAffairs, 2002), 122–23.

32   Mendelsohn, *The Martyrs*, 5; Payne, *I've Got the Light of Freedom*, 37; FBI Report, 17.

33   Dittmer, *Local People*, 53–54; FBI Report, 17.

34   Williams, *Medgar Evers*, 120; Payne, *I've Got the Light of Freedom*, 37; Wilkins, *Standing Fast*, 222.

35   Medgar Evers, 1955 Annual Report, Mississippi State Office of the NAACP, box 2, Medgar Evers Papers, MDAH.

36   Payne, *I've Got the Light of Freedom*, 51; Mendelsohn, *The Martyrs*, 8–9.

37   Roy Wilkins to Medgar Evers, May 26, 1955, box 2, Medgar Evers Papers, MDAH; "NAACP Posts Reward in Lee's Death," Jackson *State-Times*, June 2, 1955.

38   Mendelsohn, *The Martyrs*, 8–9; "NAACP Posts Reward in Lee's Death."

39   J. Todd Moye, *Let the People Decide: Black Freedom and White Resistance Movements in Sunflower County, Mississippi, 1945–1986* (Chapel Hill: University of North Carolina Press, 2004), 80.

40   Mendelsohn, *The Martyrs*, 10. See also Umoja, *We Will Shoot Back*, 34.

41   Anderson, *Separate but Equal*, 134.

42   Payne, *I've Got the Light of Freedom*, 37–38, 138–39.

43   "Is Mississippi Hushing Up a Lynching? Mississippi Gunmen Take Life of Militant Negro Minister," *Jet*, n.d. 1955, Medgar Evers Papers, MDAH; Carl M. Cannon, "Emmett Till and the Dark Path to August 28, 1963," *Real Clear Politics*, August 28, 2013, www.realclearpolitics.com/articles /2013/08/28/emmett_till_and_the_dark_path_to_aug_28_1963_119750 .html. Accessed October 31, 2016.

44   Umoja, *We Will Shoot Back*, 34; Payne, *I've Got the Light of Freedom*, 139.

45   Mendelsohn, *The Martyrs*, 11–12.

46   FBI 1956 memo quoted in Klopfer, "FBI Investigated George Lee's Murder."

47   Courts testimony, 532–33; Jay Milner, "Wounded Negro's Wife Says Mate Was 'Threatened,'" *Jackson Clarion-Ledger*, November 11, 1955, clipping in box 3 folder 6, Ed King Papers, Coleman Library, Tougaloo College (now housed at MDAH).

48   Price, "Associated Press Writer Views Mississippi Problem."

49   Courts testimony, 533–34.

50   Courts testimony, 559. See also "Is Mississippi Hushing Up a Lynching?"; Payne, *I've Got the Light of Freedom*, 37–39.

51   Gus Courts to Medgar Evers, n.d., box 2, Medgar Evers Papers, MDAH.

52   Milner, "Wounded Negro's Wife Says Mate Was 'Threatened.'"

53 Courts testimony, 533.

54 Milner, "Wounded Negro's Wife Says Mate Was 'Threatened.'" See also Caro, *Master of the Senate*, 700.

55 Payne, *I've Got the Light of Freedom*, 39.

56 James Featherston, "Two More White Men Charged in Brookhaven for Murder of Negro," *Jackson Daily News*, August 17, 1955; James Featherston, "Links Shooting of Negro with Voting Irregularities," *Jackson Daily News*, August 14, 1955. For the gunshot in his mouth, see "Negro Involved in Election Slain in Courthouse Fracas—Brookhaven Negro Linked in Election Slain by Gun Fire," *Jackson Daily News*, August 13, 1955; FBI Report, 18.

57 Arrington High, *The Eagle Eye: The Women's Voice*, 2.36, August 20, 1955, in Medgar Evers Papers, box 2, folder 7, MDAH.

58 "See Further Drop in Negro Voting in Next Tuesday's Election Following Slaying of Brookhaven Negro Leader," *Jackson Advocate*, August 20, 1955.

59 Carter, *The South Strikes Back*, 119.

60 "Lincoln County Grand Jury Unable to Get One Witness to Testify," *Jackson Advocate*, September 24, 1955.

61 Payne, *I've Got the Light of Freedom*, 40.

62 Mendelsohn, *The Martyrs*, 19.

63 Amzie Moore to James Kizart, December 20, 1955, box 1, folder 2, Amzie Moore Papers, Wisconsin Historical Society.

64 Payne, *I've Got the Light of Freedom*, 44; Charles McLaurin, interview with the author, 2012, Timothy B. Tyson Papers, Southern Historical Collection, University of North Carolina at Chapel Hill.

65 Courts testimony, 532. See also Aaron Henry with Constance Curry, *The Fire Ever Burning* (Jackson: University Press of Mississippi, 2000), 97.

66 Susan Klopfer et al., *Where Rebels Roost: Mississippi Civil Rights Revisited*, Lulu.com, 2005, 287.

12: FIXED OPINIONS

1 Whitaker, "A Case Study," 126.

2 Lloyd L. General, "Moses Wright Made His Decision—Became Hero," *Atlanta Daily World*, October 2, 1955.

3 Moses Newsom, "Emmett's Kin Hang On to Harvest Crop," *Chicago Defender*, September 17, 1955.

4 Denley, "Kinsman Recalls Tragic Night on Eve of Trial," *Chicago Defender*, September 24, 1955, 2.

NOTES

5   General, "Moses Wright Made His Decision—Became Hero."

6   James Featherston, "White 'Deplores' Slaying in Note to NAACP Which Is Creating National Issue," *Jackson Daily News*, September 1, 1955, 1.

7   *Greenwood Morning Star*, September 3, 1955, 1.

8   Beito and Beito, *Black Maverick*, 119.

9   Anderson, *Emmett Till*, 53; Whitaker, "A Case Study," 147–48.

10 *Delta Democrat-Times*, December 29, 1955, 1.

11 Dan Wakefield, "Justice in Sumner," *Nation*, October 1, 1955, in Metress, ed., *The Lynching of Emmett Till*, 120–24.

12 Mace, "Regional Identities," 240–41.

13 Bryant, Interview.

14 Shoshtak, "Crosby Smith: Forgotten Witness to a Mississippi Nightmare," 323.

15 Jack Telfer, Program Coordinator, United Packinghouse Workers Association–CIO, September 21, 1955, to Richard Durham, National Program Director, box 396, folder 7, United Packinghouse Workers of America Papers, Wisconsin Historical Society.

16 James Featherston, "State Will Not Seek Death Penalty," *Jackson Daily News*, September 19, 1955, 14.

17 Quoted in Craig Flournoy, "Reporting the Movement in Black and White: The Emmett Till Lynching and the Montgomery Bus Boycott," PhD diss., Louisiana State University, 2003, 84.

18 L. Alex Wilson, "Jim Crow Press at Trial," *Chicago Defender*, September 24, 1955, 1.

19 "Newspapers Over State Blast Murder of Negro," *Jackson Daily News*, September 3, 1955, 1; Whitaker, "A Case Study," 119–21; Beito and Beito, *Black Maverick*, 118–19.

20 Colin, "Mother's Tears Greet Son Who Died a Martyr," *Chicago Defender*, September 10, 1955, 1.

21 James Featherston, "White Orders Investigation in Slaying of Delta Negro—White 'Deplores' Slaying in Note to NAACP Which Is Creating a National Issue," *Jackson Daily News*, September 1, 1955, 1.

22 *Greenwood Morning Star*, September 1, 1955, 1.

23 "Body of Negro Found in River," *Jackson Clarion-Ledger*, September 1, 1955, 1; Featherston, "White Orders Investigation In Slaying of Delta Negro," *Jackson Daily News*, September 1, 1955, 1.

24 Beito and Beito, *Black Maverick*, 118–19.

25 Quoted in Houck and Grindy, *Emmett Till and the Mississippi Press*, 51.

26 Whitaker, "A Case Study," 132; Bryant, Interview; *Memphis Commercial Appeal*, September 4, 1955, in Metress, ed., *The Lynching of Emmett Till*, 36.

27 *Jackson Daily News*, September 5, 1955, 1, in Metress, ed., *The Lynching of Emmett Till*, 38.

28 Anderson, *Emmett Till*, 57–58.

29 *Greenwood Morning Star*, September 6, 1955, 1.

30 *Delta Democrat-Times*, September 6, 1955, in Metress, ed., *The Lynching of Emmett Till*, 38–39.

31 Houck and Grindy, *Emmett Till and the Mississippi Press*, 66.

32 Beito and Beito, *Black Maverick*, 119.

33 James Hicks, "Mississippi Jungle Law Frees Slayers of Child," *Cleveland Call and Post*, October 1, 1955, in Metress, ed.,*The Lynching of Emmett Till*, 111–13.

34 Harry Marsh, "Judge Swango Is Good Promoter for South," *Delta Democrat-Times*, September 21, 1955, in Metress, ed., *The Lynching of Emmett Till*, 60.

35 Murray Kempton, "Heart of Darkness," *New York Post*, September 21, 1955, in Metress, ed., *The Lynching of Emmett Till*, 64.

36 "Roman Circus," *Jackson Daily News*, September 22, 1955, in Metress, ed., *The Lynching of Emmett Till*, 60; "Cast in Compelling Courtroom Drama Matches Movies," *Jackson Daily News*, September 20, 1955, 1.

37 L. Alex Wilson, "Jim Crow Press at Trial," *Chicago Defender*, September 24, 1955, 1.

38 "Summary Fact Sheet of the Emmett Till Lynching Case," box 69, folder 7, 2, United Packinghouse Workers of America Papers, Wisconsin Historical Society.

39 Harry Marsh interview transcript in Ann Marie Tabb, "Perspectives in Journalism: Covering the Emmett Till Trial," honors thesis, University of Southern Mississippi, 2001, 17–18.

40 Harry Marsh, "Judge Swango Is Good Promoter for South," *Delta Democrat-Times*, September 21, 1955, in Metress, ed., *The Lynching of Emmett Till*, 61.

41 Simeon Booker, *Shocking the Conscience: A Reporter's Account of the Civil Rights Movement* (Jackson: University Press of Mississippi, 2013), 74.

42 Sam Johnson, "Two White Men Go On Trial Monday for Slaying of Negro," *Jackson Daily News*, September 18, 1955, 4.

43   "Two White Men Go On Trial," *Arkansas Gazette*, September 20, 1955, 14-B.

44   L. Alex Wilson, "Jim Crow Press at Till Trial; Frisk Newsmen; Picking of Jury Delays Opening," *Chicago Defender*, September 24, 1955, 1.

45   "Two White Men Go On Trial," *Arkansas Gazette*, September 20, 1955, 14-B.

46   Whitaker, "A Case Study," 139.

47   Rob Hall, "Lynched Boy's Mother Sees Jurymen Picked," *Daily Worker*, September 21, 1955, 1.

48   *Greenwood Morning Star*, September 25, 1955, 1; *Memphis Commercial Appeal*, September 21, 1955, in Metress, ed., *The Lynching of Emmett Till*, 59.

49   The jurors' rate of pay is in Mace, "Regional Identities," 259. The hotel details and the evidence of jury tampering by the Citizens' Council may be found in Whitaker, "A Case Study," 154, and FBI Report, 16–17.

50   Kempton, "Heart of Darkness," *New York Post*, September 21, 1955, in Metress, ed., *The Lynching of Emmett Till*, 63.

51   Whitaker, "A Case Study," 145–46.

52   *Chicago Tribune*, September 18, 1955, 1, reported: "Among the white residents of Tallahatchie County, a random sampling of public opinion produced no one who expects the two murder defendants to be found guilty." See also *Baltimore Afro-American*, October 2, 1955, 4, which reported that its reporters "could not find a single person in the Mississippi Delta who believed that the two men would be found guilty."

53   *Greenwood Morning Star*, September 23, 1955, 6.

54   W. C. Shoemaker, "Sumner Citizens Turn Public Relations Experts While Sunlight Beams At Them," *Jackson Daily News*, September 20, 1955, 1; Mace, "Regional Identities," 182.

55   Whitaker, "A Case Study," 148; Metress, ed., *The Lynching of Emmett Till*, 44–45.

56   Harry Marsh, "Radio Station Making Much Ado with Emmett Till Murder Trial," *Delta Democrat-Times*, September 22, 1955.

57   Caro, *Master of the Senate*, 701–9.

58   Harry Marsh, "Communist Writer at Trial Lauds Citizens," *Delta Democrat-Times*, September 28, 1955, 1.

59   Harry Marsh, "Radio Stations Making Much Ado with Emmett Till Murder Trial," 17. See also Tabb, "Perspectives in Journalism," 28.

60   Robert Elliot, "A Report on the Till Case: All the Witnesses Fled," *Chicago*

(November 1955): 54–55; Kevin Grimm, "Color and Credibility: Eisenhower, the United States Information Agency, and Race," MA thesis, Ohio University, 2008, 72–73 and 83.

61 Caro, *Master of the Senate*, 709.

62 Whitaker, "A Case Study," 149.

63 Dan Wakefield, writing for the *Nation*, quoted in Craig Flournoy, "Reporting the Movement in Black and White: The Emmett Till Lynching and the Montgomery Bus Boycott," PhD diss., Louisiana State University, 2003, 84.

64 David Halberstam, *The Fifties* (New York: Villard Books, 1993), 453.

65 For most white men carrying guns, see "UPWA Carries Fight for Justice to the Scene of Trial," *Packinghouse Worker* 14.1 (September 1955): 1, box 369, folder 7, United Packinghouse Workers Papers, Wisconsin Historical Society; quote is from Matthew Nichter, Chapter 3 draft, "'Did Emmett Till Die in Vain? Labor Says No': The United Packinghouse Workers of America and Civil Rights Unionism in the Mid-1950s," 11–12, in "Rethinking the Origins of the Civil Rights Movement: Radicals, Repression, and the Black Freedom Struggle," PhD diss., University of Wisconsin–Madison, 2014, in possession of the author. Many thanks to Nichter for sharing this with me.

66 James Featherston, "Till Murder Trial," *Jackson Daily News*, September 20, 1955, 7.

67 Moore interview with Michael Garvey.

68 Matthew Nichter, "'Did Emmett Till Die in Vain?,'" 13–17.

69 "UPWA Carries Fight for Justice to the Scene of Trial."

70 Mrs. Lillian Pittman, president of the Ladies Auxiliary, United Packinghouse Workers of America–CIO Local 1167, Gramercy, Louisiana, report on the Till trial, in *Packinghouse Worker* 14.9 (September 1955): 3.

71 Ibid.

72 Ibid.

73 "UPWA Carries Fight for Justice to the Scene of Trial."

74 Jack Telfer, Program Coordinator, United Packinghouse Workers of America–CIO, Sumner, Mississippi, September 21, 1955, to Richard Durham, National Program Coordinator, UPWA-CIO, box 369, folder 7, UPWA Papers, Wisconsin Historical Society.

75 "UPWA Carries Fight for Justice to the Scene of Trial."

76 Nichter, "'Did Emmett Till Die in Vain?,'" 26.

77 *Chicago Defender*, September 24, 1955, in Metress, ed., *The Lynching of Emmett Till*, 48–50; "Jim Crow Press Table," *Chicago Defender*, October 1, 1955, 3; *Jackson Daily News*, September 20, 1955, 6; James Hicks, "White Reporters Double-Crossed Probers Seeking Lost Witnesses," *Cleveland Call and Post*, October 15, 1955, in Metress, ed., *The Lynching of Emmett Till*, 161–67.

78 Kempton, "Heart of Darkness," *New York Post*, September 21, 1955, in Metress, ed., *The Lynching of Emmett Till*, 62–64.

79 Charles E. Cobb, Jr., *This Nonviolent Stuff'll Get You Killed* (New York: Basic Books, 2014), 132. This greeting was reported, with certain variations for time of day, by several observers. See, for example, Kempton, "Heart of Darkness," in Metress, ed., *The Lynching of Emmett Till*, 62–64.

80 Harvey Young, "A New Fear Unknown to Me: Emmett Till's Influence and the Black Panther Party," *Southern Quarterly* 45.4 (2008): 30.

13: MISSISSIPPI UNDERGROUND

1 Beito and Beito, *Black Maverick*, 120–21. See also Booker, *Black Man's America*, 167–68.

2 James Hicks, "Sheriff Kept Key Witness in Jail During Trial," *Cleveland Call and Post*, October 8, 1955, in Metress, ed., *The Lynching of Emmett Till*, 155–61.

3 Whitaker, "A Case Study," 150.

4 Evers, *For Us, the Living*, 172.

5 Beito and Beito, *Black Maverick*, 120–23.

6 Hicks, "White Reporters Double-Crossed Probers Seeking Lost Witnesses," in Metress, ed., *The Lynching of Emmett Till*, 164–66.

7 Beito and Beito, *Black Maverick*, 122–23.

8 Hall, "Lynched Boy's Mother Sees Jurymen Picked."

9 Featherston, "Till Murder Trial." See also *Boston Globe*, September 20, 1955.

10 Hall, "Lynched Boy's Mother Sees Jurymen Picked."

11 "Congressman Diggs, Till's Mother in Attendance at Sumner Trial," *Jackson Advocate*, September 24, 1955.

12 James Hicks, "Awakenings," *Eyes on the Prize*, transcript, www.pbs.org/wgbh/amex/eyesontheprize/about/pt_101_.html. Accessed October 31, 2016. See also Mace, "Regional Identities and Racial Messages," 384; Cobb, *This Nonviolence Stuff'll Get You Killed*, 132.

13 Hall, "Lynched Boy's Mother Sees Jurymen Picked."

14 "Roman Circus," in Metress, ed., *The Lynching of Emmett Till*, 60.

15 Featherston, "Till Murder Trial"; Anderson, *Emmett Till*, 100, 101–3.

16 Simeon Booker, "A Negro Reporter at the Till Trial" (1956), *Nieman Reports*, Winter 1999–Spring 2000, 136–37; Kempton, "Heart of Darkness," in Metress, ed., *The Lynching of Emmett Till*, 62–64.

17 Anderson, *Emmett Till*, 105.

18 Booker, "A Negro Reporter at the Till Trial"; Beito and Beito, *Black Maverick*, 123–24; James Hicks, "The Mississippi Lynching Story: Luring Terrorized Witnesses from the Plantation Was Toughest Job," *Cleveland Call and Post*, October 22, 1955, in Metress, ed., *The Lynching of Emmett Till*, 168–70; "Defender Writer in Witness Hunt," *Chicago Defender*, October 1, 1955.

14: "THERE HE IS"

1 *Delta Democrat-Times*, September 20, 1955. *Greenwood Morning Star*, September 21, 1955, estimates the courtroom crowd at "more than 400." This seems a little high. For the atmosphere, see Herbers, "Jury Selection Reveals Death Demand Unlikely," 45–53.

2 Bill Minor, *Eyes on Mississippi* (Jackson, MS: J. Prichard Morris, 2001), 191. For the church fans, see Kempton, "Heart of Darkness," in Metress, ed., *The Lynching of Emmett Till*, 62–64.

3 *New York Times*, September 22, 1955; Moses Wright, "How I Escaped from Mississippi," *Jet*, October 13, 1955.

4 Moses Wright testimony, trial transcript, 11–12. This statement has often been rendered as "Dar he," originating with an interview of James Hicks in *Eyes on the Prize: Awakenings*, almost twenty-five years later. The transcript and all of the other contemporary accounts of the trial instead report "There he is." See also Sam Johnson, "Uncle of Till's Identifies Pair of Men Who Abducted Chicago Negro," *Greenwood Commonwealth*, September 21, 1955, in Metress, ed., *The Lynching of Emmett Till*, 68; Rob Hall, "Acquittal in the Till Murder Shows Federal Intervention Is Needed," *Daily Worker*, October 2, 1955.

5 James Featherston, "Dim Light Casts Some Doubt on the Identity of Till's Abductors," *Jackson Daily News*, September 22, 1955; Booker, *Shocking the Conscience*, 74.

6 Moses Wright testimony, trial transcript, 17–18; Featherston, "Dim Light Casts Some Doubt on the Identity of Till's Abductors."

7 Murray Kempton, "He Went All the Way," *New York Post*, September 22, 1955, in Metress, ed., *The Lynching of Emmett Till*, 65.

8 Moses Wright testimony, trial transcript, 18–24.

9 Ibid., 27; Kempton, "He Went All the Way," in Metress, ed., *The Lynching of Emmett Till*, 66.

10 Moses Wright testimony, trial transcript, 32–66.

11 Wakefield, "Justice in Sumner," in Metress, ed., *The Lynching of Emmett Till*, 120–24.

12 Chester Miller testimony, trial transcript, 66–67.

13 Ibid., 67–83, 94–102.

14 Kempton, "Heart of Darkness," in Metress, ed., *The Lynching of Emmett Till*, 64.

15 Robert Hodges testimony, trial transcript, 103–12; B. L. Mims testimony, trial transcript, 113–18.

16 George Smith testimony, trial transcript, 122–33.

17 John Ed Cothran testimony, trial transcript, 139–46.

18 Ibid., 159–71.

19 "UPWA Carries Fight for Justice to the Scene of Trial," 1.

20 *New York Times*, September 23, 1955; Murray Kempton, "The Future," *New York Post*, September 23, 1955, in Metress, ed., *The Lynching of Emmett Till*, 84; *Greenwood Commonwealth*, September 22, 1955.

21 Mamie Bradley testimony, trial transcript, 185–87.

22 "Mother Insulted on Witness Stand," *Washington Afro-American*, September 24, 1955, in Metress, ed., *The Lynching of Emmett Till*, 83–84.

23 Mamie Bradley testimony, trial transcript, 188–90.

24 *Pittsburgh Post-Gazette* and *Pittsburgh Evening Bulletin*, quoted in Mace, "Regional Identities and Racial Messages," 147–48; Mamie Bradley testimony, trial transcript, 189.

25 Mamie Bradley testimony, trial transcript, 190–91.

26 Ibid., 192–209.

15: EVERY LAST ANGLO-SAXON ONE OF YOU

1 Bryant, interview. Given that Judge Swango ordered all witnesses kept out of the courtroom until after their testimony, Bryant's recollection of her sympathy may be faulty. Court officers likely did not enforce this restraint with Bryant, however, because she was a last-minute surprise witness.

2 "Till Witness Starts Guarded Life Here," clipping, n.d., box 369, folder 7, United Packinghouse Workers Papers, Wisconsin Historical Society. Clipping is probably from *Packinghouse Worker*, c. October 1955.

3 Willie Reed testimony, trial transcript, 213–25.

4 *Baltimore Afro-American*, October 1, 1955.

5 Willie Reed, Amanda Bradley, and Add Reed testimony, trial transcript, 226–59.

6 James Hicks, "Youth Puts Milam in Till Death Barn," *Washington Afro-American*, September 24, 1955, in Metress, ed., *The Lynching of Emmett Till*, 87–88.

7 Whitaker, "A Case Study," 150–51.

8 Carolyn Bryant testimony, trial transcript, 268.

9 Ibid., 261–79.

10 L. B. Otken testimony, trial transcript, 296–304.

11 H. C. Strider testimony, trial transcript, 284–95.

12 Trial transcript, 350–51.

13 James Hicks, "Called Lynch-Murder 'Morally, Legally' Wrong," *Cleveland Call and Post*, October 1, 1955, in Metress, ed., *The Lynching of Emmett Till*, 102.

14 James Featherston, "Bryant, Milam Still in Custody of Law to Face Kidnapping Charges in Leflore County after Acquittal of Murder," *Jackson Daily News*, September 24, 1955.

15 Hicks, "Called Lynch-Murder 'Morally, Legally' Wrong," in Metress, ed., *The Lynching of Emmett Till*, 102–3.

16 Whitaker, "A Case Study," 153.

17 Hicks, "Called Lynch-Murder 'Morally, Legally' Wrong," in Metress, ed., *The Lynching of Emmett Till*, 104.

18 Wakefield, "Justice in Sumner," in Metress, ed., *The Lynching of Emmett Till*, 3.

19 Sam Johnson, "Jury Hears Defense and Prosecution Arguments as Testimony Ends in Kidnap-Slaying Case," *Greenwood Commonwealth*, September 23, 1955.

20 James Featherston and W. C. Shoemaker, "Verdict Awaited in Till Trial—State Demands Conviction; Defense Says No Proof Presented, Asks Acquittal," *Jackson Daily News*, September 23, 1955.

21 Murray Kempton, "2 Face Trial as 'Whistle' Kidnappers—Due to Post Bond and Go Home," *New York Post*, September 25, 1955, in Metress, ed., *The Lynching of Emmett Till*, 107–11.

22 Till-Mobley and Benson, *Death of Innocence*, 188.

23 Featherston, "Bryant, Milam Still in Custody of Law to Face Kidnap Charges in Leflore County after Acquittal of Murder in Sumner."

24 Till-Mobley and Benson, *Death of Innocence*, 188.

25 Johnson, "Jury Hears Defense and Prosecution Arguments as Testimony Ends in Kidnap-Slaying Case"; Sam Johnson, "District Attorney Not Concerned by Outside Agitation and Pressure," *Greenwood Commonwealth*, September 23, 1955.

26 Jack Telfer to Richard Durham, box 369, folder 7, United Packinghouse Workers of America Papers, Wisconsin Historical Society, September 21, 1955, 3.

27 Kempton, "2 Face Trial As 'Whistle' Kidnappers," in Metress, ed., *The Lynching of Emmett Till*, 108.

28 Ibid., 107–11.

29 John Herbers, "Not Guilty Verdict in Wolf Whistle Murder," *Greenwood Morning Star*, September 24, 1955.

30 Kempton, "2 Face Trial As 'Whistle' Kidnappers," in Metress, ed., *The Lynching of Emmett Till*, 107–11.

31 Till-Mobley and Benson, *Death of Innocence*, 189; "Till Witness Starts Guarded Life Here."

16: THE VERDICT OF THE WORLD

1 *Arkansas Gazette*, September 2, 1955. See also *Chicago Defender*, September 10, 1955.

2 All of the major antilynching organizations agreed in 1940 that lynching was a murder by a group acting in service to race, justice, or tradition. See Christopher Waldrep, "War of Words: The Controversy over the Definition of Lynching, 1899–1940," *Journal of Southern History* 46.1 (2000): 75–100.

3 Amy Louise Wood, *Lynching and Spectacle: Witnessing Racial Violence in America, 1890–1940* (Chapel Hill: University of North Carolina Press, 2009). See James Allen, *Without Sanctuary: Lynching Photography in America* (Santa Fe: Twin Palms Publishers, 2000). See also James Allen, *Without Sanctuary: Photographs and Postcards of Lynching in America*, www.withoutsanctuary.org. Accessed October 31, 2016.

4 John Herbers, interviewed in Nelson, *The Murder of Emmett Till*.

5 James Kilgallen, "Defendants Receive Handshakes, Kisses," *Memphis Commercial Appeal*, September 24, 1955, in Metress, ed., *The Lynching of Emmett Till*, 104–7.

6 Bryant, interview; Kempton, "2 Face Trial as 'Whistle' Kidnappers," in Metress, ed., *The Lynching of Emmett Till*, 107–11; Minor, *Eyes on Mississippi*, 195.

7   Kilgallen, "Defendants Receive Handshakes, Kisses," in Metress, ed., *The Lynching of Emmett Till*, 104–7.

8   *Chicago Defender*, October 1, 1955; *Arkansas Gazette*, September 24, 1955.

9   Dittmer, *Local People*, 57; Bryant, interview. Carolyn told me years later that the marriage never recovered from the Emmett Till murder and its consequences.

10 *Chicago Tribune*, September 24, 1955.

11 Herbers, "Not Guilty Verdict in Wolf Whistle Murder."

12 *Chicago Tribune*, September 24, 1955; Whitaker, "A Case Study," 154–55; Whitfield, *A Death in the Delta*, 42.

13 *Chicago Tribune*, September 24, 1955.

14 Whitaker, "A Case Study," 155.

15 Till-Mobley and Benson, *Death of Innocence*, 189; "Till Witness Starts Guarded Life Here."

16 Herbert Shapiro, *White Violence and Black Response: From Reconstruction to Montgomery* (Amherst: University of Massachusetts Press, 1988), 411.

17 Hicks, "Mississippi Jungle Law Frees Slayers of Child," in Metress, ed., *The Lynching of Emmett Till*, 101–4.

18 Mace, "Regional Identities and Racial Messages," 153–54.

19 Johnston, *Mississippi's Defiant Years*, 36–37.

20 *Greenwood Morning Star*, September 28, 1955.

21 *Delta Democrat-Times*, September 23, 1955.

22 Mace, "Regional Identities and Racial Messages," 430.

23 *Greenwood Morning Star*, September 23, 1955.

24 Hodding Carter, "Acquittal," *Delta Democrat-Times*, September 23, 1955.

25 Hodding Carter, "Racial Crisis in the Deep South," *Saturday Evening Post* 228.25 (December 17, 1955): 26.

26 Roi Ottley, "Pathetic Mississippi," *Chicago Defender*, December 31, 1955.

27 Mrs. H. D. Schenk to the editor, *Memphis Commercial Appeal*, September 25, 1955, in Metress, ed., *The Lynching of Emmett Till*, 147–48.

28 Jason Sokol, *There Goes My Everything: White Southerners in the Age of Civil Rights, 1945–1975*, reprint edition (New York: Vintage, 2007), 40.

29 *Greenwood Morning Star*, September 30, 1955.

30 William Faulkner, open letter, in Metress, ed., *The Lynching of Emmett Till*, 42–43. See also Joel Williamson, *William Faulkner and Southern History* (New York: Oxford University Press, 1993), 303.

31 Carol Posner, *Divided Minds: Intellectuals and the Civil Rights Movement* (New York: Norton, 2001), 17. See also "Faulkner Challenged," *New York Times*, April 18, 1956; William Faulkner to W. E. B. Du Bois, April 17, 1956, in Joseph Blotner, ed., *Selected Letters of William Faulkner* (New York: Scholar Press, 1977), 398; Williamson, *William Faulkner and Southern History*, 307.

32 Fred Hobson, *But Now I See: The White Racial Conversion Narrative* (Baton Rouge: Louisiana State University Press, 1999), 129.

33 Barrow, "Here's a Picture of Emmett Till by Those Who Knew Him."

34 Office of the Honorable Adam Clayton Powell Jr., press release, October 11, 1955, in Metress, ed., *The Lynching of Emmett Till*, 133–36.

35 *New York Times*, September 25, 1955. See also Thomas Borstelmann, *The Cold War and the Color Line* (Cambridge, MA: Harvard University Press, 2003), 99.

36 A. Philip Randolph, "Call to Negro Americans," July 1, 1941, office file 93, Franklin D. Roosevelt Papers, Franklin D. Roosevelt Presidential Library. For a full scholarly examination, see Beth Tomkins Bates, *Pullman Porters and the Rise of Protest Politics in Black America* (Chapel Hill: University of North Carolina Press, 2001), especially 148–74.

37 U.S. Department of State, "Progress Report on the Employment of Colored Persons in the Department of State," March 31, 1953, box A617, group 2, NAACP Papers, Library of Congress.

38 Mary Dudziak, "Desegregation as a Cold War Imperative," *Stanford Law Review* 41 (1988): 95; Walter White, *A Man Called White* (New York: Ayer, 1948), 358–59.

39 Deyton J. Brooks to W. E. B. Du Bois, October 13, 1947, box A637, group 2, NAACP Papers, Library of Congress.

40 President's Committee on Civil Rights, *To Secure These Rights: The Report of the President's Committee on Civil Rights* (New York: Simon & Schuster, 1947), 147.

41 Dudziak, "Desegregation as a Cold War Imperative," 95.

42 Ibid., 111–12.

43 Borstelmann, *The Cold War and the Color Line*, 98–99.

44 *Baltimore Afro-American*, October 29, 1955. The writer noted newspapers in Paris, Rome, and Berlin, among others.

45 Eleanor Roosevelt, "I Think the Till Jury Will Have an Uneasy Conscience," *Memphis Press-Scimitar*, October 11, 1955, in Metress, ed., *The Lynching of Emmett Till*, 136–37.

46 Grimm, "Color and Credibility," 107.

47 Carl T. Rowan, *Breaking the Barriers: A Memoir* (New York: Little, Brown, 1991), 123.

48 U.S. State Department, *Treatment of Minorities in the United States*, 10.

49 United States Information Agency, "World Wide Press Comment on the Race Problem in the United States," April 10, 1956, quoted in Grimm, "Color and Credibility," 64–65. For a fuller discussion of larger issues around these incidents, see Mary Dudziak, *Cold War Civil Rights* (Princeton, NJ: Princeton University Press, 2000), especially 114, 118.

50 U.S. State Department, Foreign Service Dispatch, American Embassy, Brussels to Department of State, Washington, March 20, 1956, RG 59 811.411 / 4-1956, box 4158, General Records of the Department of State, National Archives, College Park, MD.

51 Grimm, "Color and Credibility," 79; *Chicago Defender*, November 5, 1955.

52 Grimm, "Color and Credibility," 72–73, 83.

53 Caro, *Master of the Senate*, 708.

54 U.S. State Department, *Treatment of Minorities in the United States*, 10.

55 "Lynching Acquittal Shocks All of Europe," *Daily Worker*, October 11, 1955. The writer quotes *Figaro* and *Le Monde*.

56 Green, *Selling the Race*, 201.

57 Quoted in "Summary Fact Sheet of the Emmett Till Lynching Case," 8, box 369, folder 7, United Packinghouse Workers of America Papers, Wisconsin Historical Society.

58 Jerry Ward and Robert J. Butler, eds., *The Richard Wright Encyclopedia* (Westport, CT: Greenwood Press, 2008), 373.

17: PROTEST POLITICS

1 "Two Rallies in North Protest Decision in Mississippi Trial," *Arkansas Gazette*, September 26, 1955, from United Press reports.

2 "6000 at Detroit Rally Protest Mississippi Verdict," *Detroit Free Press*, n.d. [September 1955], box 26, folder 4, Carl and Anne Braden Papers, Wisconsin Historical Society.

3 Whitaker, "A Case Study," 168.

4 "6000 at Detroit Rally Protest Mississippi Verdict"; Rob Hall, "Lynchers Pals Stack Jury; Protests Mount thru Land," *Daily Worker*, September 25, 1955; "Summary Fact Sheet of the Emmett Till Lynching Case," 4; "Two Rallies in North Protest Decision in Mississippi Trial." Howard's speech is reported in *Chicago Defender*, October 1, 1955; *Greenwood*

Commonwealth, September 26, 1955. The Cleveland, New Rochelle, and Newark rallies are reported in *Daily Worker*, September 26, 1955.

5   "Two Rallies in North Protest Decision in Mississippi Trial."

6   Green, *Selling the Race*, 201; *Daily Worker*, September 25, 1955.

7   Whitaker, "A Case Study," 168; Hall, "Lynchers Pals Stack Jury"; Green, *Selling the Race*, 201; Nelson, *The Murder of Emmett Till*; "NAACP Mass Meetings," *Greenwood Commonwealth*, September 26, 1955.

8   "Summary Fact Sheet of the Emmett Till Lynching Case."

9   *New York Times*, October 25, 1955.

10 *Daily Worker*, September 29, 1955.

11 "6000 at Detroit Rally Protest Mississippi Verdict." For Hurley and Marshall, see "Overflow Rally Voices Anger in Brooklyn," *Daily Worker*, October 3, 1955. See also Evers, *For Us, the Living*, 172.

12 Whitaker, "A Case Study," 168. Whitaker states that Detroit's rally drew 65,000; if this were true, the combined figure would be something like 175,000, but since several other sources say 6,000, I assume that 65,000 is a typographical error. For Detroit and New York, see William P. Jones, *The March on Washington: Jobs, Freedom, and the Forgotten History of Civil Rights* (New York: Norton, 2014), 88–89.

13 Governor Frank Clement at the 1956 Democratic National Convention, quoted in W. H. Lawrence, "Democratic Keynote Talk Assails Nixon as 'Hatchet Man' of G.O.P.; Lays 'Indifference' to President," *New York Times*, August 14, 1956.

14 For the movement to free the Scottsboro defendants, see Dan T. Carter, *Scottsboro: A Tragedy of the American South*, rev. ed. (Baton Rouge: Louisiana State University Press, 1979); James Goodman, *Stories of Scottsboro* (New York: Pantheon, 1994).

15 Rob Hall, "Set Marathon Service to Hit Till's Murder," *Daily Worker*, September 25, 1955.

16 *Chicago Tribune*, September 30, 1955.

17 "100,000 Across Nation Protest Till Lynching," *Chicago Defender*, October 8, 1955.

18 "Harlem Rally," *Chicago Defender*, October 8, 1955.

19 Robert Birchman, "10,000 Jam Till Mass Meet Here," *Chicago Defender*, October 8, 1955.

20 "Minneapolis AFL Asks U.S. Act in Till Murder," *Daily Worker*, October 3, 1955.

21 *Daily Worker*, October 3, 1955.

22 "Rally of 20,000 Here Cheers Call for Action Against Mississippi Goods," *New York Times*, October 12, 1955. See also *Daily Worker*, October 12, 1955. For the March on Washington Committees, see Jones, *The March on Washington*, 89.

23 *Daily Worker*, October 9, 1955.

24 "Donate $10,000 at Till Rally in Los Angeles," *Chicago Defender*, October 22, 1955; *Memphis Tri-State Defender*, October 22, 1955.

25 *Chicago Defender*, November 19, 1955.

26 Evers, 1955 Annual Report.

27 Huie, "Shocking Story of Approved Killing in Mississippi," 46–48.

28 "Shake-up at *Look*," *Time*, January 11, 1954; Christopher Metress, "Truth Be Told: William Bradford Huie's Emmett Till Cycle," *Southern Quarterly* 45.4 (Summer 2): 48–75. See also Flournoy, "Reporting the Movement in Black and White," 83.

29 Flournoy, "Reporting the Movement in Black and White," 83.

30 Huie, "Wolf Whistle," in Metress, ed., *The Lynching of Emmett Till*, 235.

31 William Bradford Huie interview, in Raines, *My Soul Is Rested*, 388–89.

32 Sparkman, "The Murder of Emmett Till," *Slate*, June 21, 2005.

33 Huie, "Shocking Story of Approved Killing in Mississippi," 46–48.

34 Ibid.

35 Huie, "Wolf Whistle," in Metress, ed., *The Lynching of Emmett Till*, 242.

36 Hugh Stephen Whitaker, interview by Devery S. Anderson, *Emmett Till Murder*, June 22, 2005, accessed April 3, 2016, http://www.emmetttill murder.com/new-page-92.

37 A. Philip Randolph to Martin Luther King Jr., May 7, 1956, in Clayborne Carson et al., eds., *Papers of Martin Luther King, Jr.* (Berkeley: University of California Press, 1997), 3:247–48; Taylor Branch, *Parting the Waters: America in the King Years, 1955–1963* (New York: Simon & Schuster, 1987), 209.

38 Beito and Beito, *Black Maverick*, 167–68; Carson et al., eds., *The Papers of Martin Luther King, Jr.*, 3:252–53.

39 T. R. M. Howard, speech, Madison Square Garden, May 24, 1956, tape in Chicago Public Library.

40 Ibid.

41 A. Philip Randolph, speech, Madison Square Garden, May 24, 1956, tape in Chicago Public Library; Nichter, " 'Did Emmett Till Die in Vain?,' " 31.

42   *Daily Worker*, September 10, 1955; Till-Mobley and Benson, *Death of Innocence*, "About the Authors."

43   Rose Jourdain, in Nelson, *The Murder of Emmett Till*.

44   "Clean Up Chicago First," *Greenwood Morning Star*, September 11, 1955.

45   District No. 1, United Packinghouse Workers of America, Chicago, press release, September 3, 1955, box 369, folder 7, UPWA Papers, Historical Society of Wisconsin.

46   *Daily Worker*, October 2, 1955.

47   For the importance of the Till case in stimulating the movement in the North and how those struggles built upon strategies in earlier campaigns, see Martha Biondi, *To Stand and Fight: The Struggle for Civil Rights in Postwar New York City* (Cambridge, MA: Harvard University Press, 2003), 207. The quotation from Herbert Hill came from several of our many conversations when we were colleagues in the Department of Afro-American Studies at the University of Wisconsin–Madison from 1994 until his death in 2004.

18: KILLING EMMETT TILL

1   David Blight, "Healing and History: Battlefields and the Problems of Civil War Memory," Rally on the High Ground: the National Park Service Symposium on the Civil War (Online book: National Park Service, 2001). Http://www.cr.nps.gov/history/online_books/rthg/index.htm. Accessed October 30, 2016.

2   FBI Report, 28–29, 87–91.

3   Anderson, *Emmett Till*, 333–34, 372–73, 375.

4   Willie Reed testimony, trial transcript, 223–25; *Daily Worker*, October 13, 1955; Mandy Bradley testimony, trial transcript, 253–55; Beito and Beito, *Black Maverick*, 125.

5   Autopsy report, FBI Report, 99–110; "Emmett Till," FBI Records: The Vault, accessed April 5, 2016, https://vault.fbi.gov/Emmett%20Till%20. See also Chester Miller testimony, trial transcript, 97–99; John Ed Cothran testimony, trial transcript, 150–61. For Strider's description of Till's face, see "Ask Mississippi Governor to Denounce Killing of Boy," *Chicago Tribune*, September 1, 1955.

6   E. S. Gurdjian et al., "Studies on Skull Fractures with Particular Reference to Engineering Factors," *American Journal of Surgery* 78.5 (1949):

738–39. See also Steven N. Byers, *Introduction to Forensic Anthropology* (New York: Routledge, 2016), 266–83.

7 Bryant, interview; Anderson, *Emmett Till*, 334–35.

8 FBI Report, 89–91.

9 FBI Report, 64, 89–91; Anderson, *Emmett Till*, 336, 376.

10 Martin Luther King Jr., "Letter from Birmingham City Jail," in James Melvin Washington, ed., *A Testament of Hope: The Essential Writings of Martin Luther King, Jr.* (New York: HarperOne, 1986), 295–96.

11 William H. Chafe, *The Unfinished Journey: America Since World War II* (2003; New York: Oxford University Press, 2011), 148.

12 Chester Himes, letter to the editor, *New York Post*, September 25, 1955, in Metress, ed., *The Lynching of Emmett Till*, 117.

13 William Faulkner, open letter, in Metress, ed., *The Lynching of Emmett Till*, 43. See also Williamson, *William Faulkner and Southern History*, 303.

EPILOGUE: THE CHILDREN OF EMMETT TILL

1 Stephanie Kingsley, "'So Much to Remember': Exploring the Rosa Parks Papers at the Library of Congress," *Perspectives on History: The Newsmagazine of the American Historical Association*, April 2015, accessed June 21, 2016. https://www.historians.org/publications-and-directories/perspectives-on-history/april-2015/so-much-to-remember.

2 Beito and Beito, *Black Maverick*, 139; interview with Rosa Parks by Blackside, Inc., November 14, 1985, for *Eyes on the Prize: America's Civil Rights Years (1954–1975)*, Washington University Film and Media Archive, Henry Hampton Collection, accessed April 10, 2016, http://digital.wustl.edu/cgi/t/text/text-idx?c=eop;cc=eop;rgn=main;view=text;idno=par0015.0895.080.

3 Jeanne Theoharis, *The Rebellious Life of Mrs. Rosa Parks* (Boston: Beacon, 2013), 45, 62.

4 Mace, "Regional Identities and Racial Messages," 19–22.

5 Raylawni Branch, interview by Kim Adams, October 25, 1993, transcript, 21, Oral History Collection, University of Southern Mississippi.

6 Payne, *I've Got the Light of Freedom*, 54.

7 Charles McDew, in Cheryl Lynn Greenberg, *A Circle of Trust: Remembering SNCC* (New Brunswick, NJ: Rutgers University Press, 1998), 68.

8 Bond, "The Media and the Movement," 26–27.

9 Fay Bellamy Powell, in Holsaert et al., *Hands on the Freedom Plow*, 475.

10 William H. Chafe, *Civilities and Civil Rights: Greensboro, North Carolina and the Black Struggle for Freedom* (New York: Oxford University Press, 1980), 71–72; Clayborne Carson, *In Struggle: SNCC and the Black Awakening of the 1960s* (Cambridge, MA: Harvard University Press, 1981), 9–12.

11 Frederick Harris, "Will Ferguson Be a Moment or a Movement?," *Washington Post*, August 22, 2014.

12 John Lewis discusses the powerful effect of the Till lynching on him in his memoir, *Walking with the Wind: A Memoir of the Movement* (New York: Simon & Schuster, 1998), 57–58.

13 For a full text of the speech, see Lynn Sweet, "Attorney General Eric Holder Remembers Chicago's Emmett Till," *Chicago Sun-Times*, November 17, 2014.

14 Jerome Hudson, "Eric Holder Compares Michael Brown to Emmett Till: 'The Struggle Goes On,' " *Daily Surge*, November 18, 2014, accessed April 5, 2016, dailysurge.com/2014/11/eric-holder-compares-michael-brown -emmett-till-struggle-goes.

15 Milan Kundera, *The Book of Laughter and Forgetting* (New York: Penguin, 1981), 22.

16 Stephen Kantrowitz, "America's Long History of Racial Fear," *We're History*, June 15, 2015, accessed March 15, 2016, www.werehistory.org/ racial-fear.

17 FBI, "Ten Sentenced in Hate Crime Case: Murdered Man among Multiple Victims," press release, June 16, 2015, accessed April 10, 2016, https:// www.fbi.gov/news/stories/2015/june/ten-sentenced-in-hate-crime-case/ten -sentenced-in-hate-crime-case.

18 Kim Severinson, "Weighing Race and Hate in a Mississippi Killing," *New York Times*, August 22, 2011.

19 Maya Angelou, *The Complete Collected Poems* (New York: Random House, 1994), 241.

20 Ta-Nehisi Coates, *Between the World and Me* (New York: Spiegel & Grau, 2015), 17–18.

21 Michelle Alexander, *The New Jim Crow: Mass Incarceration in an Era of Colorblindness* (New York: New Press, 2010), 6–7, *passim*.

22 Martin Luther King Jr., "Who Speaks for the South?," *Liberation* 2 (March 1958): 13–14.

23 James Baldwin, *The Fire Next Time*, in Toni Morrison, ed., *James Baldwin: Collected Essays* (New York: Library of America, 1998), 334.

24 James Baldwin, "As Much Truth as One Can Bear," in Randall Kenan, ed., *James Baldwin: The Cross of Redemption: Uncollected Writings* (New York: Pantheon, 2010), 34.

25 William J. Barber II with Barbara Zelter, *Forward Together: A Moral Message for the Nation* (St. Louis: Chalice Press, 2014).

BIBLIOGRAPHY

ARCHIVAL COLLECTIONS

Carl and Anne Braden Papers, Wisconsin Historical Society.

Citizens' Council Papers, University of Southern Mississippi.

Coleman Papers, Mississippi Department of Archives and History.

Gov. J. P. Coleman Papers, Mississippi Department of Archives and History.

Medgar Evers Papers, Mississippi Department of Archives and History.

Erle Johnston Papers, University of Southern Mississippi.

Ed King Collection, University of Southern Mississippi.

Ed King Papers, Tougaloo College (now housed at the Mississippi Department of Archives and History).

McCain Papers, University of Southern Mississippi.

Amzie Moore Papers, Wisconsin Historical Society.

National Association for the Advancement of Colored People Papers, Library of Congress.

United Packinghouse Workers of America Papers, Wisconsin Historical Society.

United States Department of State Records, United States National Archives and Records Administration, College Park, Maryland.

SELECT PRIMARY SOURCES

"Mrs. Roy Bryant, 9-2-55," lawyers' notes. Copy in possession of the author.

Carolyn Bryant Donham with Marsha Bryant, "More Than a Wolf Whistle: The Story of Carolyn Bryant Donham," unpublished memoir, Timothy B. Tyson

Papers, Southern Historical Collection, University of North Carolina, Chapel Hill.

Chicago Demographics in 1950 mmap.jpg. https://commons.m.wikimedia.org /wiki/File:Chicago Demographics in 1950 Map.jpg.

"Testimony of Rev. Gus Courts, Belzoni, Miss.," U.S. Senate, *Civil Rights—1957: Hearings before the Subcommittee on Constitutional Rights of the Committee on the Judiciary*, Eighty-fifth Congress, first session. Washington, D.C.: United States Printing Office, 1957.

Federal Bureau of Investigation Prosecutive Report on _____. https:// vault.fbi.gov/Emmett%20Till%20/Emmett%20Till%20Part%2001%20 of%2002/view.

Foreign Service Dispatch, American Embassy, Brussels to Department of State, Washington, March 20, 1956, United States Department of State RG 59 811.411/4-1956, Box 4158, General Records of the Department of State, National Archives.

T. R. M. Howard and A. Philip Randolph speeches at Madison Square Garden, May 24, 1956. Audiotape in the Chicago Public Library.

Martin Luther King Jr., "Letter from Birmingham Jail," in *A Testament of Hope: The Essential Writings of Martin Luther King, Jr.* James Washington, ed. New York: HarperOne, 1986, 295–96.

Mississippi State Sovereignty Commission online files.

Papers of Martin Luther King, Jr. Berkeley: University of California Press, 1997, 3. Clayborne Carson et al., eds.

"Summary Fact Sheet of the Emmett Till Lynching Case," box 369, folder 7, United Packinghouse Workers Papers, Wisconsin Historical Society.

"Treatment of Minorities in the United States: Impact on Our Foreign Relations, Part A Summary Review," United States State Department RG 59 811.411/4-1956, Box 4158, General Records of the Department of State, National Archives.

State of Mississippi v. J.W. Milam and Roy Bryant, In the Circuit Court Second District of Tallahatchie County, Seventeenth Judicial District, State of Mississippi, September Term, 1955, transcript. Available as Appendix A in Federal Bureau of Investigation Prosecutive Report on _____. https:// vault.fbi.gov/Emmett%20Till%20/Emmett%20Till%20Part%2001%20 of%2002/view.

The Trumbull Park Homes Disturbances: A Chronological Report, August 4, 1953 to June 30, 1955. Chicago, Mayor's Commission on Human Relations, 1955. Chicago Public Library.

NEWSPAPERS AND MAGAZINES

Amsterdam News
Arkansas Gazette
Atlanta Constitution
Atlanta Daily World
Baltimore Afro-American
Birmingham News
Boston Globe
California Eagle
Chicago
Chicago Defender
Chicago Jewish Forum
Chicago Tribune
Citizen (www.citizenscouncils.com)
Clarksdale Press Register
Cleveland Call and Post
Crisis
Daily Worker
Delta Democrat-Times
Detroit Free Press
Eagle Eye: The Woman's Voice
Ebony
Greenwood Commonwealth
Greenwood Morning Star
Grenada Daily Sentinel Star
Hattiesburg American
Jet
Jackson Advocate

Jackson Clarion-Ledger
Jackson Daily News
Jackson Free Press
Jackson State-Times
Le Democrate
Liberation
Memphis Commercial Appeal
Memphis Press-Scimitar
Memphis Tri-State Defender
Nation
Neshoba News
New York Herald Tribune
New York Post
New York Times
Nieman Reports
Packinghouse Worker
Pittsburgh Courier
Pittsburgh Evening Bulletin
Pittsburgh Post-Gazette
St. Louis Argus
Saturday Evening Post
La Sentinelle
Sumner Sentinel
Time
Vicksburg Post
Villager
Washington Post

BOOKS

Alexander, Michelle. *The New Jim Crow: Mass Incarceration in an Era of Colorblindness*. New York: New Press, 2010.

Anderson, Devery. *Emmett Till: The Murder That Shocked the World and Propelled the Civil Rights Movement*. Jackson: University Press of Mississippi, 2015.

Anderson, Henry Clay. *Separate but Equal: The Mississippi Photographs of Henry Clay Anderson*. New York: PublicAffairs, 2002.

Asch, Chris Myers. *The Senator and the Sharecropper*. New York: New Press, 2008.

Baldwin, Davarian. *Chicago's New Negroes: Modernity, the Great Migration, and Black Urban Life*. Chapel Hill: University of North Carolina Press, 2007.

Baldwin, James. *The Fire Next Time*. 1962; reprinted in Toni Morrison, ed., *James Baldwin: Collected Essays*. New York: Library of America, 1998.

Baldwin, James. Randall Kenan, ed. *The Cross of Redemption: Uncollected Writings*. New York: Pantheon, 2010.

Bartley, Numan. *The Rise of Massive Resistance*. Baton Rouge: Louisiana State University Press, 1969.

Bates, Beth Tomkins. *Pullman Porters and the Rise of Black Protest Politics in America*. Chapel Hill: University of North Carolina Press, 2001.

Beito, David, and Linda Royster Beito. *Black Maverick: T.R.M. Howard's Fight for Civil Rights and Economic Power*. Urbana: University of Illinois Press, 2009.

Biondi, Martha. *To Stand and Fight: The Struggle for Civil Rights in Postwar New York City*. Cambridge, MA: Harvard University Press, 2003.

Blackwell, Unita. *Barefootin': Life Lessons from the Road to Freedom*. New York: Crown, 2006.

Bolton, Charles. *The Hardest Deal of All: The Battle Over School Integration in Mississippi, 1870–1980*. Jackson: University Press of Mississippi, 2007.

Booker, Simeon. *Black Man's America*. New York: Prentice-Hall, 1964.

———. *Shocking the Conscience: A Reporter's Account of the Civil Rights Movement*. Jackson: University Press of Mississippi, 2013.

Borstelmann, Thomas. *The Cold War and the Color Line: American Race Relations in the Global Arena*. Cambridge, MA: Harvard University Press, 2001.

Branch, Taylor. *Parting the Waters: America in the King Years, 1955–1963*. New York: Simon & Schuster, 1987.

Brinkley, Douglas. *Rosa Parks*. New York: Penguin, 2000.

Brown, Frank London. *Trumbull Park*. 1959; Lebanon, NH: University Press of New England, 2005.

Brownell, Herbert. *Advising Ike: The Memoirs of Attorney General Herbert Brownell*. Lawrence: University Press of Kansas, 1993.

Byers, Steven N. *Introduction to Forensic Anthropology*. New York: Routledge, 2016.

Caro, Robert. *Master of the Senate: The Years of Lyndon Johnson III*. New York: Knopf, 2002.

Carson, Clayborne. *In Struggle: SNCC and the Black Awakening of the 1960s*. Cambridge, MA: Harvard University Press, 1981.

Carson, Clayborne, et al., eds. *Eyes on the Prize: America's Civil Rights Years: A Reader and Guide*. New York: Penguin, 1987.

Carter, Dan T. *Scottsboro: A Tragedy of the American South*. Baton Rouge: Louisiana State University Press, 1976.

Carter, Hodding, II. *Southern Legacy*. Baton Rouge: Louisiana State University Press, 1950.

Carter, Hodding, III. *The South Strikes Back*. Garden City, NY: Doubleday, 1959.

Cash, W. J. *The Mind of the South*. New York: Knopf, 1941.

Cecelski, David S., and Timothy B. Tyson, eds., *Democracy Betrayed: The Wilmington Race Riot and Its Legacy*. Chapel Hill: University of North Carolina Press, 1998.

Chafe, William H. *Civilities and Civil Rights: Greensboro, North Carolina and the Black Struggle for Equality*. New York: Oxford University Press, 1979.

————. *The Unfinished Journey: America Since World War II*. 7th ed. New York: Oxford University Press, 2010.

Classen, Steven D. *Watching Jim Crow: The Struggles Over Mississippi TV, 1955–1969*. Durham, NC: Duke University Press, 2004.

Coates, Ta-Nehisi. *Between the World and Me*. New York: Spiegel & Grau, 2015.

Cobb, Charles. *This Nonviolent Stuff'll Get You Killed: How Guns Made the Civil Rights Movement Possible*. New York: Basic Books, 2014.

Cobb, James. *The Most Southern Place on Earth: The Mississippi Delta and the Roots of Regional Identity*. New York: Oxford University Press, 1992.

Cohen, Adam, and Elizabeth Taylor. *American Pharaoh: Richard J. Daley: His Battle for Chicago and the Nation*. New York: Little, Brown, 2000.

Crespino, Joseph. *In Search of Another Country: Mississippi and the Conservative Counterrevolution*. Princeton, NJ: Princeton University Press, 2007.

Cronon, Edmund David. *Black Moses: The Story of Marcus Garvey and the Universal Negro Improvement Association*. Madison: University of Wisconsin Press, 1955.

Crosby, Emilye, et al., eds. *Civil Rights History from the Ground Up: Local Struggles, a National Movement*. Athens: University of Georgia Press, 2011.

Curry, Constance, et al., eds. *Deep in Our Hearts: Nine White Women in the Freedom Movement*. Athens: University of Georgia Press, 2000.

Dallek, Robert. *Lone Star Rising: Lyndon Johnson and His Times, 1908–1960*. New York: Oxford University Press, 1991.

Daniel, Pete. *Lost Revolutions: The South in the 1950s*. Chapel Hill: University of North Carolina Press, 2000.

Davies, David R. *The Press and Race: Mississippi Journalists Confront the Movement*. Jackson: University Press of Mississippi, 2001.

Dittmer, John. *Local People: The Struggle for Civil Rights in Mississippi*. Urbana: University of Illinois Press, 1994.

Dower, John W. *War Without Mercy: Race and Power in the Pacific War*. New York: Pantheon, 1987.

Drake, St. Clair, and Horace Cayton. *Black Metropolis: A Study of Negro Life in a Northern City*. 1983; New York: Harcourt & Brace, 1998.

Dray, Philip. *At the Hands of Parties Unknown: The Lynching of Black America*. New York: Random House, 2002.

Dudziak, Mary. *Cold War Civil Rights*. Princeton, NJ: Princeton University Press, 2000.

Edgerton, Douglas R. *The Wars of Reconstruction: The Brief, Violent History of America's Most Progressive Era*. New York: Bloomsbury, 2014.

Edmonds, Helen G. *The Negro and Fusion Politics in North Carolina, 1894–1901*. Chapel Hill: University of North Carolina Press, 1951.

Egerton, John. *Speak Now Against the Day: The Generation Before the Civil Rights Movement*. New York: Knopf, 1994.

Ellsworth, Scott. *Death in a Promised Land: The Tulsa Race Riot of 1921*. Baton Rouge: Louisiana State University Press, 1982.

Estes, Steve, *I Am a Man! Race, Manhood and the Civil Rights Movement*. Chapel Hill: University of North Carolina Press, 2005.

Evers, Charles, and Andrew Szanton. *Have No Fear: The Charles Evers Story.* New York: Wiley, 1997.

Evers, Myrlie, with William Peters. *For Us, the Living.* Garden City, NY: Doubleday, 1967.

Evers-Williams, Myrlie, and Manning Marable, eds. *The Autobiography of Medgar Evers.* New York: Basic Books, 2005.

Farmer, James. *Lay Bare the Heart: An Autobiography of the Civil Rights Movement.* New York: Arbor House, 1985.

Faulkner, William. *Requiem for a Nun.* New York: Random House, 1951.

Faust, Drew. *James Henry Hammond and the Old South: A Design for Mastery.* Baton Rouge: Louisiana State University Press, 1985.

Forman, James. *The Making of Black Revolutionaries.* New York: Macmillan, 1972.

Frederickson, Kari. *The Dixiecrat Revolt.* Chapel Hill: University of North Carolina Press, 2001.

Goodman, James. *Stories of Scottsboro.* New York: Pantheon, 1994.

Goodwyn, Lawrence. *Breaking the Barrier: The Rise of Solidarity in Poland.* New York: Oxford University Press, 1991.

Gottschalk, David Fort. *Veiled Visions: The Atlanta Race Riot of 1906.* Chapel Hill: University of North Carolina Press, 2005.

Graham, Allison. *Framing the South: Hollywood, Television, and Race During the Civil Rights Movement.* Baltimore: Johns Hopkins University Press, 2001.

Green, Adam. *Selling the Race: Culture, Community and Black Chicago, 1940–1955.* Chicago: University of Chicago Press, 2009.

Grossman, James. *Land of Hope: Chicago, Black Southerners, and the Great Migration.* Chicago: University of Chicago Press, 1989.

Halberstam, David. *The Children.* New York: Random House, 1998.

———. *The Fifties.* New York: Villard Books, 1993.

Hall, Jacquelyn Dowd. *Revolt Against Chivalry: Jessie Daniel Ames and the Women's Campaign Against Lynching.* New York: Columbia University Press, 1993.

Hamlin, Françoise. *Crossroads at Clarksdale: The Black Freedom Struggle in the Mississippi Delta After World War II.* Chapel Hill: University of North Carolina Press, 2014.

Hampton, Henry, et al., eds. *Voices of Freedom: An Oral History of the Civil Rights Movement from the 1950s Through the 1980s.* New York: Bantam, 1990.

Handy, W. C. *Father of the Blues: An Autobiography of W.C. Handy.* Arna Wendell Bontemps, contributor. New York: Da Capo Press, 1991.

Harkey, Ira B. *The Smell of Burning Crosses*. Jacksonville, IL: Harris-Wolfe, 1967.

Henry, Aaron, with Constance Curry. *The Fire Ever Burning*. Jackson: University Press of Mississippi, 2000.

Hill, Robert A., ed. *The Marcus Garvey and Universal Negro Improvement Association Papers*. Vols. 1–11. Berkeley: University of California Press, 1983–2006, and Durham, NC: Duke University Press, 2011.

Hirsch, Arnold R. *Making the Second Ghetto: Race and Housing in Chicago, 1940–1960*. Chicago: University of Chicago Press, 1983.

Hobson, Fred. *But Now I See: The White Racial Conversion Narrative*. Baton Rouge: Louisiana State University Press, 1999.

Holsaert, Faith. *Hands on the Freedom Plow: Personal Accounts by Women in SNCC*. Urbana: University of Illinois Press, 2012.

Houck, Davis W., and Matthew Grindy. *Emmett Till and the Mississippi Press*. Jackson: University Press of Mississippi, 2010.

Hudson-Weems, Clenora. *Emmett Till: The Sacrificial Lamb of the Civil Rights Movement*. Bloomington, IN: AuthorHouse, 2006.

Johnston, Erle. *Mississippi's Defiant Years, 1953–1973*. Forest, MS: Lake Harbor, 1990.

Jones, William P. *The March on Washington: Jobs, Freedom, and the Forgotten Civil Rights Movement*. New York: Norton, 2014.

Kennedy, Randall. *Interracial Intimacies: Sex, Marriage, Identity and Adoption*. New York: Pantheon, 2003.

———. *Race, Crime and the Law*. New York: Pantheon, 1997.

King, Mary. *Freedom Song*. New York: William Morrow, 1987.

Klopfer, Susan. *Where Rebels Roost*. Raleigh, NC: Lulu, 2005.

———. *Who Killed Emmett Till?* Mt. Pleasant, IA: Author, 2010.

Kluger, Richard. *Simple Justice*. 2005; New York, Vintage, 1976.

Lentz-Smith, Adriane. *Freedom Struggles: African Americans and World War II*. Cambridge, MA: Harvard University Press, 2011.

Lewis, Andrew. *The Shadows of Youth: The Remarkable Journey of the Civil Rights Generation*. New York: Hill & Wang, 2010.

Lewis, David Levering. *King: A Critical Biography*. New York: Praeger, 1970.

Lewis, John. *Walking with the Wind: A Memoir of the Civil Rights Movement*. New York: Simon & Schuster, 1998.

Mars, Florence. *Witness in Philadelphia*. Baton Rouge: Louisiana State University Press, 1977.

Martin, Tony. *Race First: The Ideological and Organizational Struggle of*

Marcus Garvey and the Universal Negro Improvement Association. Dover, MA: Majority Press, 1976.

Mason, Gilbert R., with James Patterson Smith. *Beaches, Blood, and Ballots: A Black Doctor's Civil Rights Struggle*. Jackson: University Press of Mississippi, 2000.

Mayhew, Howard. *Racial Terror at Trumbull Park*. New York: Pioneer Press, 1954.

McGuire, Danielle. *At the Dark End of the Street: Black Women, Rape, and Resistance*. New York: Knopf, 2010.

McMillen, Neil R. *Citizens' Council: Organized Resistance to the Second Reconstruction, 1954–1964*. Urbana: University of Illinois Press, 1971.

———. *Dark Journey: Black Mississippians in the Age of Jim Crow*. Urbana: University of Illinois Press, 1990.

Mendelsohn, Jack. *The Martyrs: Sixteen Who Gave Their Lives for Racial Justice*. New York: Harper & Row, 1966.

Metress, Christopher, ed. *The Lynching of Emmett Till: A Documentary Narrative*. Charlottesville: University of Virginia Press, 2002.

Minor, Bill. *Eyes on Mississippi*. Jackson, MS: J. Prichard Morris, 2001.

Morris, Aldon. *The Origins of the Civil Rights Movement*. New York: Free Press, 1984.

Morrow, E. Frederic. *Black Man in the White House*. New York: McFadden, 1963.

Moye, Todd. *Let the People Decide: Black Freedom and White Resistance Movements in Sunflower County, Mississippi, 1945–1986*. Chapel Hill: University of North Carolina Press, 2004.

Murch, Donna Jean. *Living for the City*. Chapel Hill: University of North Carolina Press, 2010.

Nossiter, Adam. *Of Long Memory: Mississippi and the Murder of Medgar Evers*. New York: Addison-Wesley, 1994.

Oshinsky, David. *Worse than Slavery: Parchman Farm and the Ordeal of Jim Crow America*. New York: Free Press, 1996.

Payne, Charles M. *I've Got the Light of Freedom: The Organizing Tradition and the Mississippi Freedom Struggle*. Berkeley: University of California Press, 1995.

Posgrove, Carol. *Divided Minds: Intellectuals and the Civil Rights Movement*. New York: Norton, 2001.

Raines, Howell. *My Soul Is Rested: The Story of the Civil Rights Movement in the Deep South*. New York: Putnam, 1977.

Roberts, Gene, and Hank Klibanoff. *The Race Beat: The Press, the Civil Rights Movement, and the Awakening of a Nation.* New York: Vintage, 2004.

Robnett, Brenda. *How Long? How Long? African American Women in the Struggle for Civil Rights.* New York: Oxford University Press, 1997.

Rogers, Kim Lacy. *Life and Death in the Delta: African American Narratives of Violence, Resilience, and Social Change.* New York: Palgrave Macmillan, 2006.

Rollinson, Mary G. *Grassroots Garveyism: The Universal Negro Improvement Association in the Rural South, 1920–1927.* Chapel Hill: University of North Carolina Press, 2007.

Rowan, Carl. *Breaking the Barriers: A Memoir.* New York: Little, Brown, 1991.

Salvatore, Nick. *Singing in a Strange Land: C. L. Franklin, the Black Church, and the Transformation of America.* New York: Little, Brown, 2005.

Satter, Beryl. *Family Properties: How the Struggle Over Race and Real Estate Transformed Chicago and Urban America.* New York: Picador, 2010.

Sellers, Cleveland. *The River of No Return.* Jackson: University Press of Mississippi, 1990.

Shapiro, Herbert. *White Violence and Black Response: From Reconstruction to Montgomery.* Amherst: University of Massachusetts Press, 1988.

Smith, Suzanne E. *To Serve the Living.* Cambridge, MA: Belknap Press of Harvard University Press, 2010.

Sokol, Jason. *There Goes My Everything: White Southerners in the Age of Civil Rights, 1945–1975.* New York: Knopf, 2006.

Stein, Judith. *The World of Marcus Garvey.* Baton Rouge: Louisiana State University Press, 1986.

Stokes, Gerald. *A White Hat in Argo: Family Secrets.* Lincoln, NE: iUniverse, 2004.

Swaine, Rick. *The Integration of Major League Baseball.* Jefferson, NC: McFarland, 2012.

Theoharis, Jeanne. *The Rebellious Life of Mrs. Rosa Parks.* Boston: Beacon Press, 2013.

Thompson, Julius Eric. *Percy Greene and the Jackson Advocate: The Life and Times of a Radical Black Conservative Newspaperman, 1897–1977.* Jefferson, NC: MacFarland, 1994.

Thornton, J. Wills. *Dividing Lines: Municipal Politics and the Struggle for Civil Rights in Montgomery, Birmingham, and Selma.* Tuscaloosa: University of Alabama Press, 2002.

Till-Mobley, Mamie, and Christopher Benson. *Death of Innocence: The Story of the Hate Crime That Changed America*. New York: One World/Ballantine, 2003.

Tuttle, William. *Race Riot: Chicago in the Red Summer of 1919*. New York: Atheneum, 1970.

Tyson, Timothy B. *Blood Done Sign My Name*. New York: Crown, 2004.

———. *Radio Free Dixie: Robert F. Williams and the Roots of Black Power*. Chapel Hill: University of North Carolina Press, 1999.

Umoja, Akinyele Omowale. *We Will Shoot Back: Armed Resistance in the Mississippi Freedom Movement*. New York: New York University Press, 2014.

Vollers, Maryanne. *Ghosts of Mississippi*. New York: Little, Brown, 1995.

Waldron, Ann. *Hodding Carter: The Reconstruction of a Racist*. Chapel Hill, NC: Algonquin Press, 1993.

Ward, Brian, ed. *Media, Culture, and the Modern African American Freedom Struggle*. Gainesville: University of Florida Press, 2001.

Ward, Jason. *Defending White Democracy: The Making of a Segregationist Movement and the Remaking of Racial Politics, 1936–1965*. Chapel Hill: University of North Carolina Press, 2011.

———. *Hanging Bridge: Racial Violence and America's Civil Rights Century*. New York: Oxford University Press, 2016.

Ward, Jerry W., and Robert J. Butler, eds. *The Richard Wright Encyclopedia*. Westport, CT: Greenwood Press, 2008.

Weill, Susan. *In a Madhouse's Din: Civil Rights Coverage in Mississippi's Daily Press, 1948–1968*. New York: Praeger, 2002.

Welty, Eudora. *Delta Wedding*. 1946; New York: Houghton-Mifflin, 1991.

Werner, Craig. *Higher Ground: Stevie Wonder, Aretha Franklin, Curtis Mayfield, and the Rise and Fall of American Soul*. New York: Crown, 2004.

Whitfield, Stephen. *A Death in the Delta: The Story of Emmett Till*. Baltimore: Johns Hopkins University Press, 1987.

Wilkerson, Isabel. *The Warmth of Other Suns*. New York: Random House, 2010.

Wilkins, Roy. *Standing Fast: The Autobiography of Roy Wilkins*. New York: Viking, 1982.

Williams, Michael Vinson. *Medgar Evers: Mississippi Martyr*. Fayetteville: University of Arkansas Press, 2011.

Williamson, Joel. *William Faulkner and Southern History*. New York: Oxford University Press, 1993.

Wood, Amy Louise. *Lynching and Spectacle*. Chapel Hill: University of North Carolina Press, 2009.

Woodruff, Nan. *American Congo: The African American Freedom Struggle in the Delta*. Cambridge, MA: Harvard University Press, 2003.

Wright, Richard. *Black Boy*. 1945; reprint, New York: Harper Perennial, 1993.

———. *12 Million Black Voices*. 1941; reprint, New York: Basic Books, 2002.

Wright, Simeon, and Herb Boyd. *Simeon's Story: An Eyewitness Account of the Kidnapping of Emmett Till*. Chicago: Chicago Review Press, 2010.

Zellner, Bob. *The Wrong Side of Murder Creek*. Montgomery, AL: New South Books, 2008.

ARTICLES

Amateau, Albert. "Chelsea Woman led battle to end beach segregation in 1960." *Villager* 81, no. 11: (August 11–17, 2011).

Anderson, Devery. "A Wallet, a White Woman, and a Whistle: Fact and Fiction in Emmett Till's Encounter in Money, Mississippi." *Southern Quarterly* 45, no. 4 (2008): 10–21.

Baldwin, James. "As Much Truth as One Can Bear," in Randall Kenan, eds., *The Cross of Redemption: Uncollected Writings*. New York: Pantheon, 2010.

Beauchamp, Keith. "The Murder of Emmett Louis Till: The Spark That Started the Civil Rights Movement." http://www.black-collegian.com/african/till/2005-2nd.shtml.

Beito, David, and Linda Royster Beito. "The Grim and Overlooked Anniversary of the Murder of the Rev. George W. Lee, Civil Rights Activist." *History News Network*. hnn.us/article/11744.

———. "Why It's Unlikely the Emmett Till Murder Will Ever Be Solved." *History News Network*, April 26, 2004. http://hnn.us/articles/4853_html/.

Bond, Julian. "The Media and the Movement," in *Media, Culture and the Modern African American Freedom Struggle*. Brian Ward, ed. Gainesville: University of Florida Press, 2001.

Booker, Simeon. "A Negro Reporter at the Till Trial." *Nieman Reports* (Winter 1999–Spring 2000), reprinted from January 1956.

———. "To Be a 'Negro' Newsman—Reporting at the Till Trial." *Nieman Reports* (Fall 2011).

Boyd, Herb. "The Real Deal on Emmett Till." *Black World Today* (May 18, 2004). www.afro-netizen.com.

Cannon, Carl M. "Emmett Till and the Dark Path to Aug. 28, 1963." www.real clearpolitics.com/articles/2013/08/28/emmett_till_and_the_dark_path_to _aug_28_1963_119750.html.

Courtwright, Marguerite. "The Mob Still Rides." *Negro History Bulletin* (February 1956): 105–6.

Driskell, Jay. "Amzie Moore," in Susan Glisson, ed. *The Human Tradition in the Civil Rights Movement*. Lanham, MD: Rowman & Littlefield, 2006.

Elliott, Robert. "A Reporter on the Till Case: All the Witnesses Fled." *Chicago* (November 1955): 51–56.

Feldstein, Ruth. "'I Wanted the Whole World to See': Race, Gender and Constructions of Motherhood in the Death of Emmett Till," in *Not June Cleaver: Women and Gender in Postwar America*. Joanne Meyerowitz, ed. Philadelphia: Temple University Press, 1993, 263–303.

Graham, Allison. "Civil Rights, Films, and the New Red Menace," in Brian Ward, ed., *Media, Culture and the Modern African American Freedom Struggle*. Gainesville: University of Florida Press, 2001.

Gurdjain, E. S., M.D., et al. "Studies on Skull Fractures with Particular Reference to Engineering Factors." *American Journal of Surgery* 78, issue 5 (November): 738–39.

Haygood, Wil. "The Man From *Jet*: Simeon Booker Not Only Covered a Tumultuous Era, He Lived It." *Washington Post*, July 15, 2007, W20.

Huie, William Bradford. "Shocking Story of Approved Killing in Mississippi." *Look* 20 (January 24, 1956): 46–48.

———. "What Happened to the Emmett Till Killers?" *Look*, January 22, 1957, 63–68.

———. "Wolf Whistle." Excerpted in Christopher Metress, ed., *The Lynching of Emmett Till: A Documentary Narrative*. Urbana: University of Illinois Press, 2002.

Kantrowitz, Stephen. "America's Long History of Racial Fear." *We're History*, June 15, 2015. www.werehistory.org/racial-fear.

King, Martin Luther, Jr. "Who Speaks for the South?" *Liberation* 2 (March 1958): 13–14.

Kingsley, Stephanie. "'So Much to Remember': Exploring the Rosa Parks Papers at the Library of Congress." *Perspectives on History: The Newsmagazine of the American Historical Association*, April 2015.

Kirby, Jack Temple. "The Southern Exodus, 1910–1960: A Primer for Historians," *Journal of Southern History* 49, no. 4 (November 1983): 585–600.

Luce, Phillip. "Down in Mississippi—The White Citizens' Council." *Chicago Jewish Forum*, c. 1958.

Metress, Christopher. "Truth Be Told: William Bradford Huie's Emmett Till Cycle." *Southern Quarterly* 45, no. 4 (Summer 2008): 48–75.

Oliver, Paul. *Conversations with the Blues*. Cambridge, MA: Cambridge University Press, 1997.

Panzer, Mary. "H. C. Anderson and the Civil Rights Struggle," in H. C. Anderson, *Separate but Equal: The Mississippi Photographs of Henry Clay Anderson*. New York: PublicAffairs, 2002.

Shostak, David. "Crosby Smith: Forgotten Witness to a Mississippi Nightmare." *Negro History Bulletin*, December 1974, 320–25.

Southern Poverty Law Center. "Martyrs Remembered: George Lee." www .splcenter.org/GeorgeLee.

Tisdale, John R. "Different Assignments, Different Perspectives: How Reporters Reconstruct the Emmett Till Civil Rights Murder Trial," *Oral History Review* 29, no. 1 (Winter/Spring 2002): 39–58.

Turner, Ronald. "Remembering Emmett Till." *Howard Library Journal*, 1994–1995.

Waldrep, Christopher. "War of Words: The Controversy Over the Definition of Lynching, 1899–1940." *Journal of Southern History* 66, no. 1 (February 2000): 75–100.

Ward, Bob. "William Bradford Huie Paid for Their Sins." *Writer's Digest*, September 1974, 16–22.

Whitaker, Hugh Stephen. "A Case Study in Southern Justice: The Murder and Trial of Emmett Till." *Rhetoric and Public Affairs* 8, no. 2 (2005): 189–224.

Young, Harvey. "A New Fear Known to Me: Emmett Till's Influence and the Black Panther Party." *Southern Quarterly* 45, no. 4 (2008): 22–47.

Zheng, Jianqing. "A Guided Tour Through Hell." *Southern Quarterly* 45, no. 4 (2008): 118.

DISSERTATIONS AND THESES

Berrey, Stephen Andrew. "Against the Law: Violence, Crime, State Repression, and Black Resistance in Jim Crow Mississippi." PhD diss., University of Texas at Austin, 2006.

Chamberlain, Daphne Rochelle. "'And a Little Child Shall Lead The Way': Children's Participation in the Jackson, Mississippi Black Freedom Struggle, 1946–1970." PhD diss., University of Mississippi, 2009.

Clark, Wayne Addison. "An Analysis of the Relationship Between Anti-Communism and Segregationist Thought in the Deep South, 1948–1964." PhD diss., University of Wisconsin–Madison, 1976.

Flournoy, Craig. "Reporting the Movement in Black and White: The Emmett Till Lynching and the Montgomery Bus Boycott." PhD diss., Louisiana State University, 2003.

Frazier, E. Franklin. "The Negro Family in Chicago." PhD diss., University of Chicago, 1932.

Grimm, Kevin. "Color and Credibility: Eisenhower, the United States Information Agency, and Race, 1955–1957." MA thesis, Ohio University, 2008.

Luce, Phillip Abbott. "The Mississippi White Citizens' Council, 1954–1959." MA thesis, Ohio State University, 1960, in Citizens' Council Papers, University of Southern Mississippi.

Mace, Darryl Christopher. "Regional Identities and Racial Messages: The Print Media's Stories of Emmett Till." PhD diss., Temple University, 2007.

Nichter, Matthew. "Rethinking the Origins of the Civil Rights Movement: Radicals, Repression, and the Black Freedom Struggle." PhD diss., University of Wisconsin–Madison, 2014.

Schweinitz, Rebecca Lyn. "If We Could Change the World: Children, Childhood, and African American Civil Rights Politics." PhD diss., University of Virginia, 2004.

Tabb, Ann Marie. "Perspectives in Journalism: Covering the Emmett Till Trial." Honors thesis, University of Southern Mississippi, 2001.

Whitaker, Hugh Stephen. "A Case Study in Southern Justice: The Emmett Till Case." MA thesis, Florida State University, 1963.

PAMPHLETS

Adams, Olive Arnold. *Time Bomb: Mississippi Exposed and the Full Story of Emmett Till*. 1956. Excerpted in Christopher Metress, ed., *The Lynching of Emmett Till: A Documentary Narrative*. Charlottesville: University of Virginia Press, 2002, 213–24.

Brady, Tom P. *Black Monday: Segregation or Amalgamation: America Has Its Choice*. Winona, MS: Association of Citizens' Councils, 1955.

Margaret Price. *The Negro Voter in the South*. Atlanta: Southern Regional Council, 1957.

NAACP. *M is for Mississippi and Murder*. NAACP, 1955.

Southern Regional Council. *Pro-Segregation Groups in the South*. Atlanta: Southern Regional Council, 1956. Citizens' Council Papers, University of Southern Mississippi.

FILMS

Beauchamp, Keith, director. *The Untold Story of Emmett Louis Till*. ThinkFilm and Till Freedom Come Productions, 2005.

Nelson, Stanley, producer and director. *The Murder of Emmett Till*. Firelight Media, 2003, transcript at www.pbs.org/wgbh/amex/till/filmmore/pt.html.

UNPUBLISHED WORKS

Donham, Carolyn Bryant, written by Marsha Bryant. "More than a Wolf Whistle." Southern Historical Collection, University of North Carolina at Chapel Hill.

Driskell, Jay. "Amzie Moore." Unpublished paper in possession of the author.

Nichter, Matthew. "'Did Emmett Till Die in Vain? Organized Labor Says No!': The United Packinghouse Workers of America and Civil Rights Unionism in the Mid-1950s." Unpublished paper in possession of the author.

INTERVIEWS

Branch, Raylawni, with Kim Adams, October 25, 1993, University of Southern Mississippi Oral History Collection, Hattiesburg.

Caudill, Dr., with Erle Johnston, University of Southern Mississippi Oral History Collection, Hattiesburg.

Donham, Carolyn Bryant, with Timothy B. Tyson, July 2008.

Henry, Aaron, with John Dittmer and John Jones, 1981. Mississippi Department of Archives and History, Jackson, Mississippi.

Herbers, John, in Stanley Nelson, dir., *The Murder of Emmett Till*, transcript at www.pbs.org/wgbh/amex/till/filmmore/pt.html.

Hicks, James, in "Awakenings," *Eyes on the Prize: America's Civil Rights Year*. James Hicks, transcript at www.pbs.org/wgbh/amex/eyesontheprize/about/pt_101_.html.

Howard, T. R. M., with Sidney Roger, February 26, 1958. https://archive.org/details/T.r.m.HowardInterviewBySidneyRogerFebruary261956.

Huie, William Bradford, in Howell Raines, ed., *My Soul Is Rested: Movement Days in the Deep South Remembered*. New York: Penguin, 1977, 388–89.

Mars, Florence, in University of Southern Mississippi Oral History Collection, Hattiesburg.

McDew, Charles, in Cheryl Lynn Greenberg, *A Circle of Trust: Remembering SNCC*. New Brunswick, NJ: Rutgers University Press, 1998.

McLaurin, Charles, with Timothy B. Tyson, July 22, 2013.

Moore, Amzie, with Michael Garvey, March 29 and April 13, 1977, University of Southern Mississippi Oral History Project, Hattiesburg.

———, by Blackside, Inc., 1979, for *Eyes on the Prize: America's Civil Rights Years (1954–1975)*, Washington University Film and Media Archive, Henry Hampton Collection, accessed April 1, 2016, http://digital.wustl.edu/cgi/t/text/text-idx?c=eop;cc=eop;rgn=main;view=text;idno=moo0015.0109.072.

———, in *Voices of Freedom*, Henry Hampton et al., eds. New York: Bantam, 1990.

———, in Howell Raines, ed., *My Soul Is Rested: The Story of the Civil Rights Movement in the Deep South*. New York: Putnam, 1977, 233.

Mooty, Rayfield, with Elizabeth Balanoff. Oral History in Labor Project, www.roosevelt.edu/Library/Locations/UniversityArchives/OralHistory.aspx.

Parks, Rosa, by Blackside, Inc., November 14, 1985, for *Eyes on the Prize: America's Civil Rights Years (1954–1975)*, Washington University Film and Media Archive, Henry Hampton Collection, accessed April 10, 2016, http://digital.wustl.edu/cgi/t/text/text-idx?c=eop;cc=eop;rgn=main;view=text;idno=par0015.0895.080.

Patterson, Robert, in Howell Raines, ed., *My Soul Is Rested: The Story of the Civil Rights Movement in the Deep South*. New York: Putnam, 1977, 298.

Whitaker, Hugh Stephen, with Devery Anderson, www.emmetttillmurder.com.

INDEX

283

ABOUT THE AUTHOR

TIMOTHY B. TYSON is Senior Research Scholar at the Center for Documentary Studies at Duke University and Visiting Professor of American Christianity and Southern Culture at Duke Divinity School. He also holds a position in the American Studies Department at the University of North Carolina at Chapel Hill. He taught in the Afro-American Studies Department at the University of Wisconsin–Madison from 1994 until 2004. His book *Blood Done Sign My Name* was a finalist for the 2005 National Book Critics Circle Award and won the Southern Book Critics Circle Award, the Grawemeyer Award from Louisville Presbyterian Theological Seminary, and the Christopher Award. It became a feature film in 2010. His 1999 *Radio Free Dixie: Robert F. Williams and the Roots of Black Power* won the James A. Rawley Prize and the Frederick Jackson Turner Prize from the Organization of American Historians. Tyson is the coeditor with David Cecelski of *Democracy Betrayed: The Wilmington Race Riot of 1898 and Its Legacy*, which won the 1999 Outstanding Book Award from the Gustavus Myers Center for the Study of Bigotry and Human Rights. He is a North Carolina native, a graduate of Emory University, and received his PhD from Duke University in 1994. He serves on the executive board of the North Carolina NAACP and the advisory board of the University of North Carolina Center for Civil Rights.